T0189884

THE BEDFORD SERIES IN HISTORY AND CULTURE

My Lai

A Brief History with Documents

Related Titles in
THE BEDFORD SERIES IN HISTORY AND CULTURE
Advisory Editors: Lynn Hunt, *University of California, Los Angeles*
David W. Blight, *Yale University*
Bonnie G. Smith, *Rutgers University*
Natalie Zemon Davis, *Princeton University*
Ernest R. May, *Harvard University*

My Lai

A Brief History with Documents

James S. Olson

Sam Houston State University

and

Randy Roberts

Purdue University

Palgrave Macmillan

To Marjie, who knows why.   RR

For Bedford/St. Martin's

President and Publisher: Charles H. Christensen
General Manager and Associate Publisher: Joan E. Feinberg
History Editor: Katherine E. Kurzman
Developmental Editor: Charisse Kiino
Managing Editor: Elizabeth M. Schaaf
Production Editor: Bridget Leahy
Copyeditor: Barbara G. Flanagan
Text Design: Claire Seng-Niemoeller
Indexer: Steve Csipke
Cover Design: Richard Emery Design
Cover Art: Ron Haeberle/*Life* Magazine. © Time Inc.
Composition: ComCom
Printing and Binding: Haddon Craftsmen, Inc.

Library of Congress Catalog Card Number: 97–74967

Copyright © 1998 by Bedford Books
Softcover reprint of the hardcover 1st edition 1998 978-0-312-17767-6

Manufactured in the United States of America.

9 8 7
n m l k

For information, write: Bedford/St. Martin's, 75 Arlington Street, Boston, MA 02116 (617-319-4000)

ISBN-13: 978–0–312–14227–8 (paperback)
ISBN 978-1-349-61754-8   ISBN 978-1-137-08625-9 (eBook)
DOI 10.1007/978-1-137-08625-9
Transferred to Digital Printing 2008

Acknowledgments

Fred Graham, "Army Lawyers Seek Way to Bring Ex-G.I.'s to Trial." Copyright © 1969 by The New York Times Co. Reprinted by Permission.

"The Great Atrocity Hunt." © 1969 by National Review, Inc. Reprinted by permission.

E. W. Kenworthy, "Resor Called to Testify About Alleged Massacre." Copyright © 1969 by The New York Times Co. Reprinted by Permission.

Samuel Koster, *Testimony to Peers Commission.* Reprinted from *Four Hours in My Lai,* by Michael Bilton and Kevin Sim. Copyright © 1992 by Michael Bilton and Kevin Sim. Used by permission of Viking Penguin, a division of Penguin Books USA Inc.

J. Anthony Lukas, "Meadlo's Home Town Regards Him as Blameless." Copyright © 1969 by The New York Times Co. Reprinted by Permission.

Richard M. Nixon, from *The Memoirs of Richard Nixon.* Reprinted by permission of Warner Books, Inc., New York, New York, U.S.A. Copyright © 1978. All rights reserved.

Foreword

The Bedford Series in History and Culture is designed so that readers can study the past as historians do.

The historian's first task is finding the evidence. Documents, letters, memoirs, interviews, pictures, movies, novels, or poems can provide facts and clues. Then the historian questions and compares the sources. There is more to do than in a courtroom, for hearsay evidence is welcome, and the historian is usually looking for answers beyond act and motive. Different views of an event may be as important as a single verdict. How a story is told may yield as much information as what it says.

Along the way the historian seeks help from other historians and perhaps from specialists in other disciplines. Finally, it is time to write, to decide on an interpretation and how to arrange the evidence for readers.

Each book in this series contains an important historical document or group of documents, each document a witness from the past and open to interpretation in different ways. The documents are combined with some element of historical narrative — an introduction or a biographical essay, for example — that provides students with an analysis of the primary source material and important background information about the world in which it was produced.

Each book in the series focuses on a specific topic within a specific historical period. Each provides a basis for lively thought and discussion about several aspects of the topic and the historian's role. Each is short enough (and inexpensive enough) to be a reasonable one-week assignment in a college course. Whether as classroom or personal reading, each book in the series provides firsthand experience of the challenge — and fun — of discovering, recreating, and interpreting the past.

Lynn Hunt
David W. Blight
Bonnie G. Smith
Natalie Zemon Davis
Ernest R. May

Preface

The idea for a book about My Lai first germinated years ago when we spoke about Vietnam at a Rotary Club luncheon. Our initial presentation remained relatively academic until we ventured into the question of morality and atrocities in general and My Lai in particular. The audience instantly polarized into two clear camps: one outraged at what happened at My Lai and in its aftermath; the other more tolerant, not condoning the massacre but trying to explain it. In subsequent years, in our classes on the Vietnam War, we have noticed that undergraduate students similarly divide themselves on the meaning of My Lai.

My Lai continues to haunt Americans because it challenges long-held notions of national virtue. Ever since Puritans first set foot on Massachusetts soil in 1629, a sense of mission has imbued American popular culture. We have viewed ourselves as fundamentally different from the rest of the world, free from the abuse of despots and the inhumanity of other humans. Throughout much of the twentieth century, we have also assumed responsibility for promoting human rights and ending human suffering around the world. My Lai continues to resonate because the atrocity raises the possibility that Americans are not so different after all, that all our protestations of virtue might just be so much delusion and propaganda.

When Bedford Books launched The Bedford Series in History and Culture, a publishing venture highlighting primary texts, we decided that a book on My Lai would be a perfect fit. We spent several weeks at the Army Crimes Records Center at Fort Belvoir, Virginia, and the National Archives Complex at College Park, Maryland, collecting the testimony of those who were at My Lai on March 16, 1968, and of those who investigated the incident and its cover-up. We have organized the documents so that students can examine the eyewitness testimony of participants and evaluate the behavior of those involved in My Lai and its cover-up. Students thus should be able to determine for themselves whether war crimes took place at My Lai, whether My Lai was an exception or the

rule in Vietnam, whether the nature of guerrilla warfare made such incidents more likely to occur, and whether the punishments meted out to
participants were justified.

NOTE ON THE SOURCES

The paper trail left by My Lai is now extensive, a singular irony considering the cover-up attempt. In an effort to reconstruct some of the narratives of the event, we spent time following that trail. The largest body
of unpublished material is at the National Archives Complex in College
Park, Maryland. Many of the military records formerly at the Federal
Records Center and National Archives branch in Suitland, Maryland,
have been moved to College Park. We particularly made use of the mul-
tivolume *Department of Army Review of the Preliminary Investigations in
the My Lai Incident* (commonly known as the *Peers Report*) and the trial
records of William Calley.

   We also examined the reports on My Lai conducted by the United
States Army's Criminal Investigation Division (commonly known as the
CID Reports). Conducted by Chief Warrant Officer Andre C.R. Feher,
these documents are essential for understanding what happened at My
Lai. Housed in the U.S. Army Crimes Records Center at Fort Belvoir, Virginia, they provide haunting reminders of the complexities and horrors
of war.

ACKNOWLEDGMENTS

We would like to express our appreciation to the archival staffs at the
Army Crimes Records Center at Fort Belvoir, Virginia, and at the National
Archives Complex at College Park, Maryland, for their skill and unfail-
ing good humor in putting up with our requests. We are also indebted to
the following scholars who offered critical appraisals of the manuscript:
Marilyn Young, New York University; Stephen Kneeshaw, College of
the Ozarks; Craig Lockard, University of Wisconsin, Green Bay; Daniel
Czitrom, Mount Holyoke College; Bruce Schulman, Boston University;
and Michael Barnhart, State University of New York, Stony Brook. We
also extend our gratitude to the following people at Bedford Books:
Charles Christensen, publisher; Joan Feinberg, associate publisher;
Katherine Kurzman, sponsoring editor; Elizabeth Schaaf, managing editor; Bridget Leahy, production editor; Barbara Flanagan, copyeditor; and

Richard Emery and Terry Govan for the cover design. Finally, we are grateful to Charisse Kiino, our developmental editor at Bedford Books, for an extraordinary job in seeing this project to completion.

James S. Olson
Randy Roberts

Contents

My Lai

A Brief History with Documents

INTRODUCTION

The Road to My Lai

For Ronald Ridenhour the war would not go away. Even though he had been discharged recently from the army and was back with his family in Phoenix, Arizona; even though he had a job and had been accepted into college to study English literature; even though outwardly his life had returned to normal—still, the memories of his time in Vietnam stayed with him. He was particularly haunted by one horrible story he had heard, nine months earlier, about a company of soldiers who had gone on a killing rampage in a tiny hamlet in Quang Ngai province. In April 1968 he had been sitting in a bar in Vietnam, drinking a beer and swapping war stories with Pfc. Charles "Butch" Gruver.[1]

"Did you hear about Pinkville?" Gruver asked, referring to a Task Force Barker operation in Quang Ngai province.

"No, what did you do at Pinkville?"

"We went in there and killed everybody."

Ridenhour had pressed Gruver for more details. With a growing sense of horror, Ridenhour listened as Gruver casually discussed the operation—how Charlie Company had gone into the village; shot anything that moved; herded old men, women, and children into groups and executed them; mowed down people and animals alike. Ridenhour was no novice to violence; he had been out in the field long enough to see civilians killed and to witness the things decent men might do in the heat of battle. But he had never heard anything like Gruver's story. He repeatedly asked Gruver if they had killed everyone, and Gruver answered, "Yes."

Later Ridenhour remembered thinking, "These no good son-of-a-bitches. Look at what they've gotten me into. Look what they've gotten me into." He believed that Gruver had handed him an ethical bomb. If

---

[1] Ridenhour's conversation with Gruver and attitudes toward the My Lai massacre are recounted in Michael Bilton and Kevin Sim, *Four Hours in My Lai* (New York: Penguin, 1992), 214–18.

the story that Gruver told was true, then Ridenhour had to either report it and turn in several friends or keep quiet and become part of the crime.

During the next several months, Ridenhour sought out conversations with several other men he knew who had been in Charlie Company at the time of the Pinkville operation. They were good men, men who Ridenhour knew would risk their lives for him. He hoped that they would just tell him that Gruver had been drunk or had exaggerated or had just plain lied. But they didn't. They confirmed Gruver's tale, adding more details and coloring in the picture in Ridenhour's mind. One soldier, a deeply religious Mormon who Ridenhour knew was decent and honest, said that he and another friend ate lunch close to where a large number of villagers had been shot. The problem was that they were not all dead. A few were suffering and crying out for help that would never arrive. So the two members of Charlie Company put down their plates, picked up their rifles, killed the wounded Vietnamese villagers, and then returned to their meals.

Back in the United States, now a civilian himself, Ridenhour could not get Pinkville out of his mind. Finally, on March 29, 1969, he wrote a letter about what he knew was true to Mo Udall, his local Democratic congressman, who had come out against the war. On April 2, 1969, he posted the letter to Udall and sent copies to about thirty other leading government officials, including President Richard Nixon, Secretary of Defense Melvin Laird, the chairman of the Joint Chiefs of Staff, and Senators Edward Kennedy, Barry Goldwater, Eugene McCarthy, and William Fulbright.

Ridenhour's letter got right to the point. "It was late in April, 1968 that I first heard of 'Pinkville' and what allegedly happened there. I received that first report with some skepticism, but in the following months I was to hear similar stories from such a wide variety of people that it became impossible for me to disbelieve that something rather dark and bloody did indeed occur sometime in March, 1968 in a village called 'Pinkville' in the Republic of Viet Nam."[2]

In forceful, precise language Ridenhour recounted exactly what he had uncovered about the Pinkville operation. He described how a four-year-old boy, suffering from an arm wound and watching in disbelief as his village was being destroyed, was killed by a burst of M-16 fire; how a Sec-

---

[2]Letter from Mr. Ronald L. Ridenhour to Secretary of Defense, March 29, 1969, in William R. Peers, *Report of the Department of the Army Review of the Preliminary Investigation into the My Lai Incident*, vol. 1, *The Report of the Investigation* (Washington, D.C.: U.S. Government Printing Office, 1970), 7–11.

ond Lieutenant "Kally" had machine-gunned at least three large groups of villagers; how somewhere between three hundred and four hundred noncombatant villagers had been killed; how Captain Ernest Medina had told one soldier who had refused to engage in the killings "not to do anything stupid like write [his] congressman." Ridenhour told the truth about what he had heard and whom he had heard it from. He implicated not only his friends but also such officers as Lieutenant "Kally" (William Calley) and Captain Ernest Medina.

Toward the end of his letter Ridenhour arrived at his central concern: "I remain irrevocably persuaded that if you and I do truly believe in the principles, of justice and the equality of every man, however humble, before the law, that form the very backbone that this country is founded on, then we must press forward a widespread and public investigation of this matter with all our combined efforts. I think that it was Winston Churchill [*sic*] who once said 'A country without a conscience is a country without a soul, and a country without a soul is a country that cannot survive.'" Ronald Ridenhour's letter set in motion an investigation not only of something that had occurred in an area designated "Pinkville" but also of the role of American soldiers and United States policy in Vietnam.

AMERICA'S LONGEST WAR

Vietnam was a French colony in 1945, but Vietnamese nationalists had long chafed under foreign domination. They had fought for centuries against Chinese imperialism. But just after they expelled the Chinese in the early nineteenth century, France imposed a new form of colonialism. French officials and their Vietnamese surrogates dominated the political bureaucracy, seized land and businesses, and imposed Roman Catholicism on the country. Vietnamese nationalists launched a guerrilla campaign against the French, but French military power was too great.

Japan then did what the Vietnamese could not do. When Germany overran France in 1940, Japan occupied much of French Indochina. Vietnamese nationalists, led by Ho Chi Minh, then turned on the Japanese. They did not want an Asian imperialist master any more than a Western one. In fact, Ho Chi Minh worked with the Office of Strategic Services, the forerunner of the Central Intelligence Agency (CIA), during World War II, providing intelligence data and helping to return American pilots shot down over the South China Sea. For its part, the United States armed Ho Chi Minh's five-thousand-man army.

As World War II came to a close, President Franklin D. Roosevelt actually contemplated liberating Vietnam from French control. The Vietnamese, he thought, were ready for independence. But France, a key United States ally, would have none of it, and Great Britain also opposed Vietnamese independence. When World War II ended, France moved back to Indochina and reimposed its colonial control of Cambodia, Laos, and Vietnam.

Ho Chi Minh and his Vietminh army resumed their guerrilla war. Within a matter of two years, whatever sympathy the United States had once had for Ho Chi Minh evaporated. He was a Communist, and he believed that only with communism could the Vietnamese ever really enjoy freedom and equality. He was also a genuine nationalist set on achieving Vietnamese independence. But in the Red scare, cold war atmosphere of the late 1940s, Ho's communism loomed larger in American eyes than his nationalism. By 1950, the United States was providing hundreds of millions of dollars to help France defeat the Vietminh.

The assistance was of little use. France steadily lost ground in Vietnam, and Ho Chi Minh recruited an army of more than 100,000 highly skilled, committed troops. In the spring of 1954, the Vietminh inflicted a humiliating defeat on French forces at Dienbienphu. With the French empire crumbling, the United States worked to preserve a non-Communist base in Vietnam. Many American policymakers believed that if Vietnam fell to communism, a chain reaction, like falling dominoes, would eventually bring all of Southeast Asia, and perhaps East Asia, under Communist control. In the Geneva Accords of 1954, the United States managed to see to the division of Vietnam at the seventeenth parallel. North Vietnam would be Communist and South Vietnam would be non-Communist. Free elections would be held in 1956 to reunite the country.

Ho Chi Minh became the leader of the Democratic Republic of Vietnam (North Vietnam), and Ngo Dinh Diem, an anti-Communist Vietnamese nationalist, became president of the Republic of Vietnam (South Vietnam). The free elections were never held. When the CIA determined that Ho Chi Minh would win handily in the north and the south, the United States canceled the elections. Ho Chi Minh began recruiting sympathetic Vietnamese into a southern army, which became known as the Vietcong. The Vietcong then launched a guerrilla war against the American-backed regime of Diem.

Diem was a corrupt, anti-Buddhist Roman Catholic who soon alienated most South Vietnamese. The Vietcong grew stronger, and in the late 1950s and early 1960s Presidents Dwight Eisenhower and John F. Kennedy deployed hundreds of U.S. military advisers to train the South

Vietnamese army. The Vietcong continued, however, to gain ground, and by September 1963 more than 16,000 U.S. military advisers were in-country. Ngo Dinh Diem was assassinated by his own generals on November 1, 1963, and three weeks later, President Kennedy was assassinated in Dallas, Texas. Vice President Lyndon B. Johnson became president of the United States.

Throughout 1964 and early 1965, South Vietnam continued to deteriorate. Political instability—complete with rebellions and coup d'états—became endemic. Johnson sent several thousand additional U.S. military advisers to South Vietnam in an attempt to stem the tide. In August 1964, North Vietnamese torpedo boats attacked an American destroyer in the Gulf of Tonkin, leading Johnson to ask Congress for permission to respond with appropriate force. Congress immediately passed the Gulf of Tonkin Resolution, on which Johnson based his order to bomb North Vietnam. By early 1965, it was clear that if the United States did not introduce regular ground troops, the Vietcong would triumph. By then Johnson knew that the war was getting out of hand, but he refused to disengage, not wanting to be "the first president to lose a war." In March 1965, Johnson deployed the first contingent of U.S. marines to Vietnam, and by the end of the year more than 184,000 American ground troops were in the country.

Despite the growing American commitment, the government of South Vietnam grew weaker and the Vietcong, now sustained by troops and supplies from North Vietnam, grew stronger. American military officials kept raising the "minimum number of troops" necessary to win the war, and by the end of 1966 more than 325,000 American soldiers were fighting in Vietnam. Late in November 1967, General William Westmoreland promised Johnson that the end of the war was in sight, that the presence of nearly 500,000 U.S. troops had finally overwhelmed the enemy. The news gratified Johnson, who wanted to have Vietnam behind him before the 1968 presidential election.

But it was not to be. On January 31, 1968, the Vietcong launched the Tet Offensive. By that time, large numbers of North Vietnamese soldiers had crossed the border to join forces with the Vietcong. They attacked U.S. and South Vietnamese forces throughout the country. U.S. troops repelled the attack and inflicted huge casualties on the Vietcong, but Tet was a political disaster at home. Vietnam appeared to be a never-ending quagmire. At the end of March, Lyndon Johnson announced that he would not run for reelection. And during the last stages of the Tet Offensive, in My Lai village of Quang Ngai province, the most controversial event of the war took place.

THE GROUND WAR IN QUANG NGAI

Some people remembered the sheer beauty of Quang Ngai province. "God, this is such a beautiful country," an American guide told novelist and travel writer Paul Theroux. In the mid-1970s, as Theroux made his way by train from Hue to Danang, he thought that "no picture could duplicate the complexity of the beauty: over there, the sun lighted a bomb scar in the forest, and next to it smoke filled the bowl in a valley; a column of rain from one fugitive cloud slanted on another slope, and the blue gave way to black green, to rice green on the flat fields of shoots, which became, after a strip of sand, an immensity of blue ocean." Theroux was prepared for the scars of war, but not for what stretched out in front of his eyes. "We were at the fringes of a bay that was green and sparkling in the sunlight. Beyond the leaping jade plates of the sea was an overhang of cliffs and the sight of a valley so large it contained sun, smoke, rain, and cloud—all at once—independent quantities of color. I had been unprepared for this beauty; it surprised and humbled me. . . . Who has mentioned the simple fact that the heights of Vietnam are places of unimaginable grandeur?"[3]

Some American soldiers who fought in Quang Ngai province noticed the scenery—the patchwork green rice paddies and white sandy beaches, the picturesque rivers and huge bamboo and banana trees. Located in northeastern South Vietnam, Quang Ngai is bordered on the west by the foothills of the Annamese Mountains and on the south by the South China Sea; viewed from almost any direction it presents a postcard-like loveliness. But far more soldiers concentrated on the dangers, for Quang Ngai was as deadly as it was beautiful. As a province, it had a rebellious, violent, and bloody record, a history of extraordinary hostility toward foreign occupation forces, whether Chinese, Japanese, French, or American. Few provinces more strenuously resisted French occupation, few more strongly resented the American-backed government in Saigon. The province was a harbor of Vietcong sentiment and a fertile region for Vietcong recruitment. Although the South Vietnamese government and its American allies controlled the area around the provincial capital, Quang Ngai City, they were the enemy in most rural regions.[4]

---

[3]Paul Theroux, *The Great Railway Bazaar: By Train through Asia* (New York: Houghton Mifflin, 1975), 258–59.

[4]Bilton and Sim, *Four Hours*, 56–57; Seymour M. Hersh, *My Lai 4: A Report on the Massacre and Its Aftermath* (New York: Random House, 1970), 3–4.

As soon as American combat troops arrived in South Vietnam in 1965, the Pentagon deployed U.S. marines to Quang Ngai to clear the region of Vietcong. During the engagement, code-named Operation Starlight, the marines claimed substantial victories, but they failed to stem Vietcong activity or dent Communist sympathy. In 1966, one senior officer explained his unit's mission: "We've been told by our superiors that in many areas there isn't any chance of pacifying the people, so instead we've got to sanitize our region—kill the Viet Cong and move the civilians out. We are not going to be able to make the people loyal to our side. So we are going to sterilize the area until we can win it back."[5]

The officer's statement revealed several crucial problems. First, such words as *sanitize* and *sterilize*—euphemisms for killing people and destroying villages—indicated that the enemy was no longer regarded as human. Communists, whether Vietcong troops or the villagers who supported them, had become something of a germ, in need of a good scrubbing and prompt elimination. Second, the statement also raised the question, If a village is "sterilized"—that is, destroyed—what is left to be won back? Increasingly, the U.S. military and civilian leaders had begun to view the war in terms of territorial conquests, not the attainment of villagers' support. Increasingly, Americans had locked on to the idea of a military victory, while the Vietcong were pursuing a political one.

Winning territory in a political war proved a thorny proposition. What did the occupation of land mean when the population of that land supported the Communist forces and when Vietcong guerrillas moved in and out of the territory? For many U.S. marines, the enemy was both difficult to identify and maddeningly elusive. The war in Vietnam was so unlike the war most soldiers expected. Many had been raised on images of John Wayne fighting World War II. Philip Caputo remarked in *A Rumor of War,* his personal war memoir, that as he considered enlisting in the marines he saw himself "charging up some distant beachhead, like John Wayne in *Sands of Iwo Jima,* and then coming home a suntanned warrior with medals on [his] chest." But Vietnam was no John Wayne movie, no heroic landings on enemy-held shores or daring charges up enemy-controlled mountains to plant the American flag. There was no *Sands of Iwo Jima* in Vietnam, and no Mount Suribachi.[6]

Instead of reliving the glorious engagements of World War II, ground troops in Vietnam spent days trudging through rice paddies, where the

---

[5]Hersh, *My Lai,* 4.

[6]Philip Caputo, *A Rumor of War* (New York: Ballantine, 1977), 6; Lawrence H. Suid, *Guts and Glory: Great American War Movies* (Reading, Mass.: Addison-Wesley, 1978), 102–9.

mud sucked at their feet, or through jungles, where heat and a fog of flies and mosquitoes made every moment uncomfortable. Occasionally, they saw the enemy and confronted him in a short firefight. More often the only indication that they were on hostile soil came when one of them was shot by an invisible sniper or stepped on a booby trap. It was an atmosphere in which frustration bred anger, and anger hatched hate—hate for the Vietcong, hate for Vietnam, hate for the Vietnamese. A common joke told by marines mapped their state of mind: "The loyal Vietnamese should all be taken and put out to sea in a raft. Everybody left in the country should then be killed, and the nation paved over with concrete, like a parking lot. Then the raft should be sunk." "In the end anybody who was still in that country was the enemy," a soldier stationed in Quang Ngai commented. "The same village you had gone in to give them medical treatment . . . you could go through that village later and get shot at on your way out by a sniper. Go back in, you wouldn't find anybody. Nobody knew anything. . . . We were trying to work with these people, they were basically doing a number on us. . . . You didn't trust them anymore. You didn't trust anybody."[7]

Racial prejudices developed quickly. Soldiers referred to Vietnamese—friends and enemies alike—as "gooks," "slopes," "slant-eyes," or "dinks." Many just gave up trying to separate friendly from unfriendly Vietnamese and considered them all enemies. Some Americans used a modified Wild West analogy, mouthing such sentiments as "The only good gook is a dead gook." In the minds of many soldiers, the Vietnamese had become some subhuman species—repulsive, duplicitous, and deadly.

Military policy reinforced those stereotypes and beliefs. The logic behind "free-fire zones" contributed to the idea that few Vietnamese could be trusted. In theory, free-fire zones were areas from which non-Communists had been removed and placed in fortified hamlets; in theory, those left behind were by definition Vietcong or Vietcong sympathizers, and they remained in the free-fire zones at their own risk. In free-fire zones American soldiers went on "search and destroy missions," and American pilots rained the land with bombs and napalm. Of course, the policy meant that innocent civilians—including old men, women, children, and infants—would be killed. But the theory that those civilians had been given an opportunity to leave made the policy morally defensible, at least in the minds of American policymakers. Since much of Quang

[7]Hersh, *My Lai*, 11; Bilton and Sim, *Four Hours*, 39–40.

Ngai province was controlled by the Vietcong, U.S. military leaders designated large sections of it as free-fire zones.

In the spring of 1967, because large numbers of North Vietnamese troops had crossed the demilitarized zone between North and South Vietnam and entered Quang Ngai, United States forces escalated the war there. Up to that point, the fighting in Quang Ngai had been handled by the marines; now the United States army entered the fray. This meant more manpower, more firepower, and more destruction. The first army operation was code-named Task Force Oregon, and its objective was to "sanitize" the entire province. In four months, U.S. military leaders claimed that they killed 3,300 Vietcong, captured 800 weapons, and arrested 5,000 suspects. Of course, there were other results as well. By the military's own estimates, between 1965 and 1967 the war rendered more than 138,000 Quang Ngai civilians homeless and led to the destruction of more than 70 percent of the homes in the province. Roads were destroyed, hospitals overflowed, and thousands of peasants swarmed into the already overcrowded Quang Ngai City.[8]

Journalist Jonathan Schell of the *New Yorker* was a witness to Task Force Oregon. "The overriding, fantastic fact that we are destroying, seemingly by inadvertence, the very country we are supposedly protecting" struck Schell as tragic. Everywhere he looked, he saw the smoke and fire of destruction, from pockmarked rice paddies to burning villages. He saw technological war on an unprecedented scale, but next to nothing that would win the "minds and hearts" of the local peasants. One private from California told Schell, "When I got here, some of the villages were wiped out, but quite a lot were still there. Then every time I went out there were a few less, and now the whole place is wiped out as far as you can see. The G.I.s are supposed to win the people's confidence, but they weren't taught any of that stuff. I went through that training, and I learned how to take my weapon apart and put it back together again, and how to shoot, but no one ever told me a thing about having to love people who look different from us and who've got an ideological orientation that's about a hundred and eighty degrees different than us. When we got here, we landed on a different planet. . . . Even when a Vietnamese guy speaks perfect English I don't know what the hell he's talking about."

---

[8]Jonathan Schell, *The Real War: The Classic Reporting on the Vietnam War* (New York: Pantheon, 1988), 197–99; Neil Sheehan, *A Bright Shining Lie: John Paul Vann and America in Vietnam* (New York: Vintage, 1988), 686–89.

Another private from Texas agreed. "The trouble is, no one sees the Vietnamese as people. They're not people. Therefore, it doesn't matter what you do to them."[9]

CHARLIE COMPANY

One of the army units destined for service in the Quang Ngai region of Vietnam was Charlie Company, 1st Battalion, 20th Infantry. As several scholars of the My Lai massacre would observe later, there was simply nothing unusual about Charlie Company. "Charlie Company was very average," noted writers Michael Bilton and Kevin Sim. Most of the men were high school graduates between the ages of eighteen and twenty-two; there was a fairly even division between black and white soldiers; and the company had the look of a cross section of American society. Also typically, the company was slightly understaffed. In theory a company is composed of four platoons (three rifle and one weapons) of forty men each. Each platoon is led by a lieutenant and is further divided into four ten-man squads, each commanded by a sergeant. By the time Charlie Company moved into My Lai, the size of its platoons had shrunk to two or three squads. But again, there was nothing out of the ordinary about such attrition.[10]

Like thousands of other American soldiers during the Vietnam War, the men who were assigned eventually to Charlie Company received their basic training in such mainland camps as Fort Benning, Fort Polk, and Fort Jackson, then shipped off to Schofield Barracks in Hawaii for their final training before going into the war zone. Their training in Hawaii was designed to teach them to follow the orders of their superiors. Following orders—quickly and without asking questions—was the glue of the system. It held everything together, allowing individual men to act decisively as a unit under the harshest conditions. Although the code of international land warfare might have been mentioned from time to time, in a vague nonspecific way, its implications were not discussed in any great detail. After all, its central tenet—"Men who take up arms against one another in public war do not cease on this account to be moral beings, responsible to one another and to God"—could weaken the glue of the system. Such codes raised thorny questions: Was there a potential conflict between following the orders of one's superiors and following the dictates of one's conscience? Where does one draw a clear line between right

---

[9]Schell, *Real War*, 191, 230.
[10]Bilton and Sim, *Four Hours*, 50–51; Hersh, *My Lai*, 21.

and wrong—a proper order and an improper one—in the blurred landscape of war?

Difficult questions, and ones to which few soldiers gave much thought. Paul Meadlo, a rifleman in Charlie Company, recalled clearly the priorities that were drilled into him during training: "From the first day we go in the service, the very first day, we all learned to take orders and not to refuse any kind of order from a noncommissioned officer. . . . An officer tells you to go and stand on your head, it's not your right to refuse that order, and you go out there and do it because you're ordered to. . . . If you refuse the order, the son-of-a-bitch might shoot you or the next day you spend the rest of your life in the stockade for refusing an order, but you're trained to take orders from the first day you go to that damned service, and you come back and, all right, you want to try some people that had to take orders."[11]

If the men in Charlie Company did not receive adequate training in the treatment of noncombatants and prisoners, they did acquire a range of other military skills. They learned how to execute an amphibious assault and how to guard each other's back as they moved through hostile territory. They drilled how to march, hold a weapon, and kill with rifle, bayonet, or bare hands. Army officials were more than satisfied with Charlie Company's progress. In fact, the unit once won the Company of the Month award.

Perhaps even more important, the men learned from their commissioned and noncommissioned officers whom to trust and whom not to trust. "Trust each other" was the message they picked up in countless ways. The mythology and often the reality of combat was that the man next to you in battle would fight to the death with you and even sacrifice his life to save yours. And you would do the same for him. In battle it was you and your unit against them—a faceless, generic enemy. In Vietnam, officers instructed their men, the enemy could be any Vietnamese. Of course, all soldiers in the North Vietnamese army (NVA) were enemies, as were the Vietcong, the South Vietnamese supporters of Ho Chi Minh and the North Vietnamese Communist regime. But beyond these obvious enemies were the thousands, perhaps millions, of South Vietnamese civilians who supported the Communists. The enemy, American soldiers were told, might be the seemingly harmless Vietnamese cleaning a camp latrine or turning hamburgers in the NCOs' club. The enemy might be the prostitute in a Saigon bar or the eight-year-old girl reaching out to a GI for candy. And the enemy, officers left no doubt, would kill an American soldier in a heartbeat and not think twice about it.

[11]Peers, *Report,* vol. 2, bk. 25, 11–12.

Journalist Michael Herr recorded the heartthrob of paranoia that pulsated through Vietnam. "One place or another," he wrote, "it was always going on, rock around the clock, we had the days and he had the nights. You could be in the most protected space in Vietnam and still know that your safety was provisional. . . . The roads were mined, the trails booby-trapped, satchel charges and grenades blew up jeeps and movie theaters, the VC got work inside all the camps as shoeshine boys and laundresses and honey-dippers, they'd starch your fatigues and burn your shit and then go home and mortar your area. Saigon and Cholon and Danang held such hostile vibes that you felt you were being dry-sniped every time someone looked at you." For hundreds of thousands of American soldiers it was clearly a world of "us" and "them."[12]

The task of instilling the necessary esprit de corps into Charlie Company fell to Captain Ernest L. Medina, a stocky, crewcutted Mexican American who had enlisted in the army to escape the poverty of his youth. He had risen through the ranks by dint of his own efforts and had a passionate faith in discipline. He never suffered fools gladly, and anyone who made a mistake was apt to receive an earful of abuse and instantly understand the origin of Medina's nickname, "Mad Dog." But he was fair and did not play favorites, and the men who served under him respected and admired his no-nonsense approach to the army. Charles West, a member of Charlie Company, said Medina was one of "the best officers I've known" and Charlie Company was "the best company ever to serve in Vietnam." The soldiers of Charlie Company, he later recalled, "operated together or not at all! We cared about each and every individual and each and every individual's problems. This is the way we were taught by Captain Medina to feel toward each other. We were like brothers."[13]

Lieutenant William L. Calley, the leader of Charlie Company's 1st Platoon, was as incompetent as Medina was competent. If a Hollywood director were going to cast a cold-blooded wartime villain, the short, nondescript Calley would not get a second glance. He had a soft face, which even during his service years seemed to border on pudgy. He had a weak chin, nervous gerbil-like eyes, and pasty skin. Certain that a draft call was imminent, Calley had joined the army, hoping that in enlisting rather than being conscripted he might have a chance at more interesting duty. Bilton and Sim aptly described Calley as "a bland young man burdened, it seemed, with as much ordinariness as any sin-

---

[12]Michael Herr, *Dispatches* (New York: Avon, 1977), 13–14.
[13]"The Massacre at Mylai," *Life,* December 5, 1969, 39.

gle individual could bear; and almost too much of the conventional and commonplace to retain what is necessary for human identity." His parents had raised him in middle-class security, he demonstrated no tendency toward anything exceptional, and, as much as any human being, he blended into the background of any situation. Most observers agreed that he was the sort of person you had to know well even to remember that you knew him at all.[14]

Because he had a couple of years of college, Calley was assigned to Officer Candidate School (OCS), but he turned out to be a poor excuse for an officer. He never mastered reading maps, had difficulty carrying out basic assignments, and had absolutely zero leadership ability. Most of his men regarded him as something of a pint-sized joke, a Napoleon want-to-be who demanded a level of respect he never earned. He reminded one of his platoon members of "a little kid trying to play war," and though he continually tried to impress Medina, the veteran officer held him in low regard. Medina called Calley "Sweetheart" and ridiculed him in front of his platoon. "Medina didn't like Calley," recalled a member of Calley's platoon. "Calley was always doing something wrong. . . . I wondered sometimes how he got through OCS; he couldn't read no darn map and a compass would confuse his ass." But by 1968 there was a shortage of second lieutenants, and Calley, even with his obvious shortcomings, somehow made the grade.[15]

In the second week of December 1967, Charlie Company was deployed to Quang Ngai province in Vietnam. The company's first month in combat passed uneventfully. Stationed to the west of Duc Pho, in the mountainous southern part of the province, Charlie Company pulled five-to-ten-day patrols, tramping through the bombed-over terrain searching for an elusive enemy. Several years before, Duc Pho had been one of the most dangerous areas in Vietnam's most dangerous province. But brutal marine firepower had largely tamed the area. Reporter Jonathan Schell estimated that in the hostile areas of Duc Pho, and especially along the strategically important Route 1, "ninety to one hundred percent of the houses . . . had been destroyed." Much of Duc Pho was a wasteland, and there was little left for Charlie Company save walking and looking. Recalling that first month, Calley later testified, "We didn't get shot at and we didn't shoot at—we took a couple of enemy. We shot at a few of them but it was very, very rare and it would be only if we came up on top of the

[14]Bilton and Sim, *Four Hours,* 49.
[15]Hersh, *My Lai,* 20–21; [William Calley Court-Martial Transcripts, Federal Records Center, Suitland, MD, 3769.]. (A full transcript of the Calley trial is in the National Archives Complex, College Park, Maryland.)

hill and we would catch them in an open valley and generally bring artillery gunships on them."[16]

In late January 1968, Charlie Company's uneventful season ended. The company was transferred to northern Quang Ngai and reassigned to a newly formed five-hundred-man strike force. The new unit, attached to the Americal Division, was designated Task Force Barker after its commander, Lieutenant Colonel Frank Barker, and its area of operation was code-named Muscatine after a town in Iowa. Any resemblance between the peaceful, bucolic town of Muscatine, Iowa, and Muscatine, Vietnam, ended with the name. Muscatine was hot; it was Vietcong territory, and neither U.S. marines, Korean marines, the South Vietnamese army, nor the U.S. army had been able to "pacify" the area. Yet that was exactly Task Force Barker's assignment.

The enemy in northern Quang Ngai was the Vietcong's 48th Local Force Battalion. The 48th Battalion had been a constant problem for the South Vietnamese and their American allies. Its soldiers were disciplined, were trained to endure severe hardships, and enjoyed the general support of the villagers in their areas of operation. Most important, they were not overly bold. Their leaders understood that patience was their greatest virtue, and they planned and executed only small-scale operations. The idea was to create problems but remain intact. Their patience, combined with their discipline, made them a dangerous and frustrating enemy for American leaders who demanded immediate results.

If Americans longed for action and a large-scale engagement, they got it in the early hours of January 31, 1968. It occurred during Tet, the lunar New Year celebration. Both sides had agreed to a Tet cease-fire, and South Vietnamese authorities had even lifted the general ban on fireworks. Shortly after midnight, laughter and celebration turned to horror and bloodshed. In a carefully planned, coordinated action, the Vietcong and the North Vietnamese army launched a countrywide assault against South Vietnam. They attacked 36 of 44 provincial capitals, 5 of 6 autonomous cities, 64 of 242 district capitals, and 50 hamlets. In Saigon, the Communists carried their attack into the courtyard of the United States embassy, and in the old imperial capital of Hue, the scene of the bloodiest fighting, the Communists took control of the citadel and the Imperial Palace of Peace.

The Tet Offensive reached Quang Ngai City, located on Route 1, just after four in the morning. Vietcong units attacked the local citadel, the airfield, a South Vietnamese training center, the provincial jail, and the Army of the Republic of Vietnam (ARVN) headquarters. They also

---

[16] *United States v. William L. Calley*, 3773.

assaulted Americal Division headquarters. Secure in their base camp outside Quang Ngai City, the soldiers of Charlie Company saw the traces of mortar shells and rockets light up the night sky; they heard the mechanized sounds of war. Far to their west, they witnessed evidence—the flashes of light and rolling thunder of sound—that Americal Division was in a heated fight. Several years later, William Calley recalled his thoughts as he looked toward the battle taking place at Americal Division headquarters: "It's just like the enemy is closing in on you.... You realize that the enemy was in such strength, that they had massed enough to take on your division elements, think what [they] could do to your company if they found you alone."[17]

The Communists had hoped that the Tet Offensive would lead to a general uprising, that the "people" would take arms against their oppressors, and that South Vietnamese soldiers would desert to join the Communist cause. It did not happen. Instead, the offensive was a massive military defeat for the Communists, especially the Vietcong, who shouldered most of the fighting and suffered most of the losses. But that verdict took months, even years, to reach. In the days and weeks after the start of the offensive, the meaning and end of Tet were still in doubt. It was a time of anxiety and second-guessing—anxiety for the American soldiers out in the field and second-guessing for journalists and many Americans at home.

It was at this historically crucial moment that Charlie Company experienced the realities of combat in Vietnam. The Tet Offensive played itself out quickly in Quang Ngai province. Massive American and South Vietnamese firepower repelled the Communist moves, and the 48th Battalion, badly crippled, retreated down the Tra Khuc Valley toward the coast, through the region designated "Pinkville" on American maps, that included My Lai.

During the days and weeks that followed Tet, Task Force Barker attempted to locate and eliminate the remainder of the 48th Battalion. U.S. and South Vietnamese authorities feared that Tet was only the first phase of the Communist offensive, and they wanted to attack the Communists before they could regroup. For the three companies of Task Force Barker—Alpha, Bravo, and Charlie—this entailed patrolling the Muscatine area and attempting to trap or engage the enemy.

It was a difficult, dangerous, and frustrating assignment, one that made February seem twice as long as any other month of the year. Charlie Company spent long stretches away from base camp, rising at dawn, patrolling through rice paddies and villages, eating K-rations, fending off

[17]Ibid., 3779.

insects, and digging in at night. Life became one long, hot, dirty, uncomfortable grind, marked by numbing stretches of boredom and brief but intense moments of jagged fear. As the month dragged on, Charlie Company began to take casualties. Several men stepped on booby traps, others were caught in short brushfires. They seldom saw any Vietcong, but they knew their enemy was out there—watching, waiting for just the right moment to attack.

For the men of Charlie Company and their commanders, February was like a final exam. Some passed. Others did not. Some accepted the discomforts, contained their fears, did their jobs, and treated Vietnamese civilians with respect. Others had trouble adjusting. Of the officers, Lieutenant Calley was the one most often singled out for criticism. One soldier called him a "glory-hungry person . . . the kind of person who would have sacrificed all of us for his own personal advancement." A second soldier noted that Calley's men hated him so much that they "put a bounty on his head." Even Captain Medina referred to Calley as "Lieutenant Shithead," a circumstance that did not make Calley's men feel overly confident. Even worse, Calley showed poor leadership qualities in the field. More than once his inability to read a map got his platoon lost, and he reacted badly under fire.[18]

As casualties mounted, Charlie Company demanded revenge. But revenge against whom? It was unusual to see a Vietcong with a weapon, let alone get a shot at one. But soldiers did see civilians—civilians in Vietcong country—and gradually the sharp line between civilians and Vietcong began to blur. As in all wars, soldiers learned from other soldiers, and myths, rumors, oft-repeated tales, and superstitions became firmly held and scientifically proven axioms. The most common belief was that any Vietnamese man, woman, or child might be a Vietcong operative. The old man with a stringy goatee, the young woman with a child in her arms, the smiling children with their hands reaching out for candy—any one or all of them might lead you toward a booby trap or throw a grenade at you when you least expected it. So expect it. Don't drop your guard for a second.

Varnado Simpson, a soldier in Charlie Company, described the mental process that turned civilians into the enemy: "Who is the enemy? How can you distinguish between the civilians and the noncivilians? The same people who come and work in the bases at daytime, they just want to shoot and kill you at nighttime. So how can you distinguish between the two? The good or the bad? All of them look the same." Many members

[18]Bilton and Sim, *Four Hours,* 70–95; Hersh, *My Lai,* 23–38.

of Charlie Company echoed Simpson's opinion. It seemed impossible to separate neutral Vietnamese from the enemy, and the safest solution seemed simply to consider anyone in enemy territory as the enemy.[19]

This attitude allowed Charlie Company to drift swiftly into a culture of violence in which anything seemed permissible. Cut off from the civilization and rules of conduct that they had learned in the United States, isolated on long patrols through Vietnamese villages, frustrated at seeing their friends and comrades killed or maimed, many soldiers adopted a new code of behavior, one that permitted the killing of prisoners, the torture of suspects, the cutting off of ears of the dead, and the rough treatment and rape of civilians. Michael Bernhardt, a product of La Salle Military Academy who believed firmly in the rules of wartime engagement, was appalled at the behavior of many members of his company. He later described how prisoners were used as "mine detectors"—forced to walk ahead of the point man during mine sweeps—and brutally kicked and punched. And in the villages some distance from Route 1, rape was not uncommon. When asked by investigators if rape was a widespread practice, Bernhardt replied, "I thought it was, sir. It was predictable. In other words, if I saw a woman, I'd say, 'Well, it won't be too long.' That's how widespread it was."[20]

Once again, leadership failed Charlie Company. The officers who should have set a better example and held their men to a higher code of conduct did not. Captain Medina, remembered Michael Bernhardt, said "everything that walked and didn't wear a uniform was VC. . . . He was as much a nut as anybody else. He was pissed off at the people and had no respect for them." With Medina providing the moral yardstick, platoon and squad leaders fell into line. On several occasions Bernhardt yelled "Dung lai," a phrase he believed meant "stop," at Vietnamese peasants. But he never got the right reaction; instead of stopping, the person he yelled at usually moved along even faster. It occurred to him that he was using the wrong phrase or voice inflection. Maybe what the Vietnamese understood him to say was "get lost." Lieutenant Calley, however, was not interested in the nuances of Vietnamese. He ordered Bernhardt to shoot anyone who did not stop.[21]

By March the mood of Charlie Company was decidedly ugly. Long patrols, gnawing discomforts, and mounting casualties had taken their toll. Morale was low, and men complained that they had been singled out

[19]Bilton and Sim, *Four Hours,* 74.
[20]Peers, *Report,* vol. 2, bk. 25, 463–64.
[21]Hersh, *My Lai,* 24–31; Peers, *Report,* vol. 2, bk. 25, 380–81.

for the worst duties. Perhaps, some soldiers in Charlie Company speculated, one of their officers was feathering his own nest at their expense. With each new frustration, with each new painful reminder of the agony of war, the level of unsanctioned violence against Vietnamese civilians increased. On March 14, Gregory Olsen witnessed a brutal attack on a female civilian by several members of Charlie Company. It happened after a booby trap killed one soldier and blew the legs off another. The men took out their frustrations by killing the first Vietnamese they saw. "Why in God's name does this have to happen?" Olsen wrote to his father. "These are all seemingly normal guys; some were friends of mine. For a while they were wild animals. It was murder, and I'm ashamed of myself for not trying to do anything about it."[22]

Frustration was not the private reserve of Charlie Company. Higher up the command level, the leaders of the 11th Brigade and the Americal Division were frustrated by their complete inability to locate and engage the Vietcong's 48th Local Force Battalion. Everyone was sure the battalion was somewhere in Quang Ngai, but nobody was certain exactly where. Slowly, however, a consensus was forming. Intelligence officers ventured that the 48th was somewhere near My Lai, a group of subhamlets of Son My village near the South China Sea. They asserted that the 48th had suffered dearly during the Tet Offensive and had taken refuge in the hills north and northwest of one of the hamlets, designated My Lai 4. There, the intelligence officers believed, the 48th was regrouping and rebuilding. This information generated one simple conclusion: a surprise assault in the area around Son My village could result in a total rout of the 48th.

This information also led to other conclusions, the most important of which was that there were few neutral civilians in Son My. Captain Eugene M. Kotouc, an intelligence officer attached to Task Force Barker, later testified: "The civilian population was known to be rather active sympathizers with the VC. The VC was a local unit. The VC came from the families. There were mothers, fathers, and sons of the VC. . . . They were not people who came down from the north." Kotouc and several other intelligence officers maintained that an early-morning assault would be best to avoid contact with civilians. According to their logic, every morning the peasants went to the local markets and would be away from the village.[23]

Behind any operation are certain assumptions—best guesses, really,

[22]Hersh, *My Lai,* 37–38.
[23]Peers, *Report,* vol. 2, bk. 16, 101–2.

based on past experiences and gathered intelligence. But often these guesses become hardened truths as they filter down the chain of command, and by the time they reach the ears of the average soldier they have become set in stone. By the time Captain Medina gathered together the company in the dying light of March 15, 1968, to go over plans for the next day's operation, he had heard and bought into the official line: The enemy—some 250 to 280 strong—was outside of My Lai 4; neutral civilians would be away at market when the assault occurred; and civilians remaining in the village would probably be Vietcong or Vietcong sympathizers.

What Medina was told to do with any civilians—and what exactly he told his men to do with them—remains somewhat vague. Different men heard different things and interpreted what they heard differently. But everyone agreed that they expected to aggressively encounter the enemy, and not much thought was given to civilians. Those who heard task force leader Lieutenant Colonel Frank Barker's orders were certain he wanted My Lai 4 leveled. Kotouc remembered, "Colonel Barker said he wanted the area cleaned out, he wanted it neutralized, and he wanted the buildings knocked down. He wanted the hootches burned, and he wanted the tunnels filled in, and then he wanted the livestock and chickens run off, killed, or destroyed." Medina also recalled that Barker had given orders to destroy the village.[24]

Medina passed on Barker's orders to Charlie Company. After explaining their mission, he told the men to expect "a hell of a good fight." He left no doubt in anyone's mind that he wanted Vietcong killed and My Lai destroyed. He wanted houses burned, livestock slaughtered, wells collapsed, and crops ruined. In short, he wanted nothing to remain that could render support for the Vietcong.

One member of Charlie Company asked about women and children. Should they be killed too? Medina later testified that he replied, "No, you do not kill women and children. You must use common sense. If they have a weapon and are trying to engage you, then you can shoot back, but you must use common sense." Gregory Olsen confirmed Medina's reply: "Captain Medina would never have given an order to kill women and children." Many soldiers in Charlie Company, however, carried away a different interpretation of Medina's instructions. For them, Medina's main message was that it was time for revenge, that they should think about all their friends who had been killed or wounded and then go into My Lai and settle some scores. "It was clearly explained that there were to be no

[24]Ibid., 105; Bilton and Sim, *Four Hours,* 95–97; Hersh, *My Lai,* 39–40.

prisoners," recalled Sergeant Kenneth Hodges. "The order that was given was to kill everyone in the village. Someone asked if that meant women and children. And the order was: everyone in the village. . . . It was quite clear that no one was to be spared in that village."[25]

Lieutenant Calley had vivid memories of Medina's briefing. He remembered that Medina mentioned every man Charlie Company had lost and then used a shovel to map out the plan of attack in the sandy soil. He also recalled that Medina emphasized speed and aggression: Move through the hamlets — My Lai 4, My Lai 5, My Lai 6 — allow no one to get behind you, and "neutralize" as you go along. Civilians, Medina explained, had been cleared out of the area. Everyone remaining was the enemy. "So it was our job to go through destroying everyone and everything in there," Calley recalled, "not letting anyone or anything get behind us and move into Pinkville."[26]

THE MY LAI MASSACRE

"We expected strong VC resistance," Dennis Conti of Charlie Company recalled thinking on the morning of March 16. "We were really expecting trouble. We were all psyched up." Most members of his company expressed the same feelings. Although the night before there had been talk over beers about killing civilians, the soldiers had thought more about engaging the enemy. Their mission, after all, was not to kill civilians; it was to clear the area of Communists. That, and the danger they were bound to face, occupied their thoughts as they sat in their helicopters en route to their landing zone.[27]

Calley was in one of the first helicopters. It was the job of his 1st Platoon to secure the landing zone, a mission that often resulted in heavy casualties. By 7:40 A.M. his platoon was on the ground, about 150 meters west of My Lai 4. The ground was wet and thick with harvest-season elephant grass. There was plenty of cover for the enemy. But the enemy was nowhere near My Lai 4. Calley's squad, loaded heavily with ammunition, landed with guns blazing, but no Vietcong guns blazed back. It was as if they had landed in the coldest hot spot in Vietnam. The only Vietnamese they saw was one man in a rice paddy, who might have been waving or

[25]Bilton and Sim, *Four Hours,* 98–101; Gregory Olsen, CID Report; Harry Stanley, CID (Criminal Investigation Division) Report, Fort Belvoir, Virginia; Hersh, *My Lai,* 40–43.
[26]*United States v. William L. Calley,* 3792–93.
[27]Peers, *Report,* vol. 1, 5–15.

might have been farming. No one knows for sure; a burst of machine gun fire cut him down too fast to tell.

With the landing zone secured, Charlie Company's 2nd and 3rd Platoons were soon on the ground. Two other companies of Task Force Barker landed at about the same time—one to the north of My Lai 4, the other to the south. The idea was for Charlie Company to move through My Lai 4 (known to the Vietnamese who lived there as Tu Cung), while the companies to the north and south trapped all fleeing Vietcong. No sooner was Charlie Company on the ground than its troops began to move toward My Lai 4. Calley's 1st Platoon and Lieutenant Stephen Brooks's 2nd Platoon led the assault; they were to move through the hamlet and eliminate the opposition. Lieutenant Jeffrey La Crosse's 3rd Platoon was held in reserve; its task was to move in later and kill the livestock and burn the hooches. On paper it was a textbook operation: move in fast, surprise the enemy, cut off escape routes, and drive them to the sea. It was like a fisherman with a net, or a group of western farmers on a jackrabbit roundup.

The problem was that there were no fish to net or jackrabbits to round up. But there were civilians—the same ones who were supposed to be at market. The soldiers moved into My Lai in small squads, firing at anything that moved. Cows, water buffalos, pigs, chickens, ducks—they were all fair game in this hunt. The men were tense with fear, edgy on adrenaline, and expecting the enemy to surface at any time. So they fired their weapons, forgetting—if they ever really learned—the rules of engagement. They shot at civilians who appeared to be running; they threw grenades into houses; they screamed orders in a mixture of English and poor Vietnamese and they expected to be obeyed.

Confusion reigned. My Lai 4 was not an open hamlet; it was divided and subdivided by bamboo trees, banana trees, and bushes. Because Calley's and Brooks's platoons were also divided and subdivided, no single person had a good view of the entire operation. For many, the constant sound of gunfire signaled that fighting was taking place somewhere. But in fact, nowhere was there any sign of the enemy. The only people shot were civilians—old men, women, and children. The later testimony of the soldiers singled out scores of horrors and brutalities. Herbert Carter, one of the men who refused to shoot civilians, remembered seeing a woman holding a baby coming out of a hut. She was crying because someone had shot her little boy. "She came out of the hut with her baby and Widmer shot her with an M16 and she fell. When she fell, she dropped the baby and then Widmer opened on the baby with his M16 and killed the baby, too."[28]

Other men witnessed or participated in similar episodes. Soldiers shot old men sitting outside their homes, women carrying water, children searching for places to hide. Some children were shot as they reached out their hands toward a GI in hope of receiving candy. Women were raped at gunpoint. Calley recalled seeing one of his men raping a woman and telling him "to get his pants back up and get over to where he was supposed to be."[29]

But Calley himself was responsible for the most horrific incidents. Most soldiers in Charlie Company who encountered civilians did not kill them; instead they herded the Vietnamese into open areas where they could guard them. This process threatened to slow down the progress of the operation, however. Several times during the morning Medina radioed Calley for progress reports. He wanted to know why Charlie Company was not moving faster. Calley replied that the civilians his men had gathered were the main problem. Medina was not in a mood for excuses. Calley later testified that Medina "told me to 'waste the Vietnamese and get my people out in line, out in the position they were supposed to be.'" Calley proceeded to follow Medina's instructions to the letter. He ordered his men to kill the unarmed civilians. Dennis Conti recalled one of the mass executions. Calley had ordered Conti and Paul Meadlo to help him kill the civilians. Conti begged off, saying that he would watch the tree line. Then Calley and Meadlo opened fire. As Conti remembered the next few moments, "Meadlo fired a while. I don't know how much he fired, a clip, I think. It might have been more. He started to cry, and he gave me his weapon. I took it, and he told me to kill them. And I said I wasn't going to kill them. At the time . . . the only thing left was children. I told Meadlo, I said: 'I'm not going to kill them. [Calley] looks like he's enjoying it. I'm going to let him do it.'"[30]

All morning long, Ronald Haeberle, an army photographer, took pictures of the killings. He snapped some of the photos with an army-issue camera and army-issue black-and-white film. With his own personal camera, Haeberle took dozens of color shots. At no time did he protest the killings or try to protect civilians. He was a journalist, exercising the obligatory professional detachment from the event.

Altogether, American soldiers killed between four hundred and five

[28]Herbert L. Carter, CID Report.
[29]*United States v. William L. Calley,* 3813.
[30]Peers, *Report,* vol. 2, bk. 24, 382.

hundred Vietnamese in My Lai 4. But not every member of Charlie Company participated in the slaughter. Some fired only when they were given direct orders to fire, others simply refused to fire at all. All the men were presented with a difficult moral choice—to follow what they believed were their orders or to do what their consciences told them was right. At one point helicopter pilot Hugh Thompson Jr., who was flying reconnaissance in support of the operation, was shocked by what he saw from his Plexiglas "bubble ship" and landed his chopper between soldiers and a group of defenseless villagers. He told the American soldiers—his countrymen—that he had his gunner protecting him so they better not shoot the villagers. He simply could not abide what he saw happening. And when he returned to base he reported what he had seen, first to a chaplain and then to Colonel Oran K. Henderson, commander of the 11th Infantry Brigade. Thompson let his superiors know that a bloody atrocity had taken place at My Lai and that American troops had been responsible.[31]

By noon the slaughter was over. What had happened in the four hours of the morning of March 16, 1968, was a tragedy. There is no doubt that the soldiers were given orders to shoot. Lieutenant William Calley, the ranking officer in My Lai, both ordered and participated in the worst executions, and he certainly believed that he was following the orders of his commander, Captain Ernest Medina. In the mass confusion of the morning—in what one military strategist has called the "fog of war"—things happened that probably no one could have predicted.

AFTERMATH

But what happened after that morning was coldly calculated. A massacre, not a battle, had taken place. The signs of indiscriminate killing of civilians were apparent. Battles mean that your own men get killed and wounded; the only casualty in Charlie Company was one accidental, self-inflicted wound. Battles successfully waged mean the capture of enemy soldiers and weapons; officially the "battle" of My Lai listed 128 enemies killed but only three weapons recovered. Looking just at the numbers, any experienced officer could have guessed what had taken place. Besides, Thompson had reported the truth. But there was no serious investigation, only a cover-up that reached up the chain of command from Captain Medina to Lieutenant Colonel Frank Barker to Colonel Oran Henderson, to Major General Samuel Koster, commander of the

---

[31]Peers, *Report,* vol. 2, bk. 8, 16–17, 37, 40, 43–46.

Americal Division. But after a brief and perfunctory investigation of Thompson's charges, military authorities concluded that no infractions of the military code of engagement had occurred.

Ronald Ridenhour's letter of March 29, 1969, opened a Pandora's box of scandal, recrimination, and anguish, inspiring military and congressional inquiries. Lieutenant General William Peers headed the official military investigation, and he eventually identified 224 serious violations of the military code. On Peers's recommendation, the army filed war crimes and obstruction of justice charges against two generals, four full colonels, four lieutenant colonels, four majors, six captains, and eight lieutenants. A few reporters had already received tips about the investigation, and on November 13, 1969, Seymour Hersh broke the story in newspapers around the country. The December 5, 1969, issue of *Life* magazine printed Ron Haeberle's grisly color photographs, which he had sold to the magazine because they were his own property. The massacre and cover-up had suddenly become a massacre, cover-up, and scandal.

The revelations triggered a storm of controversy throughout the United States. For antiwar activists, the massacre and cover-up stood as stark totems to the evil of the Vietnam War. Military spokesmen insisted that My Lai was an isolated event, not at all representative of the entire war. Many Americans offered another explanation: Vietnam was a different war, and the men of Charlie Company should not be held accountable. After all, they had been forced to engage in a conflict nobody wanted, to confront an enemy that was indistinguishable from a noncombatant, to fight a war without fronts, and to put themselves in harm's way even though the military was constrained from using all of its resources to protect them. My Lai was a tragedy, to be sure, but blame for My Lai rests in Washington, D.C., not in Charlie Company, this line of reasoning concluded.

The results of the investigations and trials were equally controversial. In the end, only William Calley was found guilty, convicted of the premeditated murder of twenty-two civilians and sentenced to life in prison at hard labor. As punishment for participating in the cover-up, Major General Koster was reduced in rank to brigadier general, and his assistant, Brigadier General George Young, was officially censured. Everyone else, including Colonel Oran K. Henderson, was acquitted or had all charges dropped.

The judicial outcomes immediately became fodder for more controversy and recrimination. In spite of what many considered to be a mountain of prima facie evidence, six-man military tribunals consistently refused to convict, except in Calley's case, where the evidence was overwhelming, egregious, and aggravated. The jurors were usually Vietnam War veterans themselves. In the air-conditioned courtrooms where they deliberated Charlie Company's fate, perhaps they considered Vietnam another world, a surrealistic realm where different standards of civilization prevailed. Or perhaps they too had seen, or had participated in, atrocities during their tours of duty. Convicting the men of Charlie Company might have seemed hypocritical. Perhaps they were just circling the wagons in defense of the military, hoping the not guilty verdicts would serve as a sort of damage control and bring closure to the controversy.

Closure, however, was something that would never come to the villagers of My Lai, the American soldiers who fought there, or the controversy it generated. What happened was brutally simple, and the events of March 16, 1968, have been reconstructed to the minute. But why it happened remains something of a mystery. What exactly did the officers who planned the My Lai assault order? What did the officers and soldiers on the ground believe was their mission? And, more importantly, how could such an event happen? The following documents suggest partial answers, but no combination of documents can tell the full story, because there was no single narrative. Rather, the day had many narratives—official, unofficial, personal. Depending on one's politics, attitude toward war, and understanding of human nature, the narrative is subject to constant revision.

1

The Official Story

During the years of the Vietnam War, public trust in American politicians reached an all-time low. Journalists coined the term "credibility gap" to describe the chasm between official descriptions of the war and the reality of the conflict. The phrase "credibility gap" was first used by reporter David Wise in a May 23, 1965, article for the *New York Herald Tribune.* At the time, there were only a few thousand American combat troops stationed in South Vietnam, but the antiwar movement was gaining momentum. Public suspicions about the truthfulness of the Johnson administration escalated as well. Murray Marder, a reporter for the *Washington Post,* popularized the term "credibility gap" in an article on December 5, 1965, in which he argued that government officials were being disingenuous in painting a rosy picture of a war that was actually going quite badly. The 1971 publication in the *New York Times* of the so-called Pentagon Papers exposed to the entire world just how disingenuous public officials had been. Four American presidents—Eisenhower, Kennedy, Johnson, and Nixon—had been guilty of misleading the American people.

My Lai was certainly no exception; in fact, it may very well be the best illustration of the "credibility gap." More than a year passed between Charlie Company's assault on My Lai on March 16, 1968, and April 2, 1969, when Ronald Ridenhour mailed his letter to prominent American political and military leaders. In his letter, Ridenhour described what he had heard about the events at My Lai. In response to the exposé, both houses of Congress and the U.S. army launched investigations of the incident. By that time, Army officials in Vietnam had already engaged in a conspiracy to cover up the truth about My Lai. During the first year after the massacre, military and civilian newspapers had treated the attack as just another successful U.S. army combat operation in Vietnam, one that resulted in a body count of 128 enemy soldiers. The following documents reveal the official story of what happened at My Lai.

JAY A. ROBERTS

Press Release

March 1968

*Sergeant Jay Roberts was a reporter with the Public Information Depart-
ment of the 11th Brigade. He accompanied Charlie Company during the
assault on My Lai and witnessed the killings, but after the incident on
March 18, 1968, he wrote a press release that completely sanitized the operation, transforming a wanton massacre into a "battle."*

Chu Lai, Vietnam (Americal IO)—For the third time in recent weeks, the
Americal Division's 11th Brigade infantrymen from Task Force Barker
raided a Viet Cong stronghold known as "Pinkville" six miles northeast
of Quang Ngai, killing 128 enemy in a running battle.

The action occurred in the coastal town of My Lai where, three weeks
earlier, another company of the brigade's Task Force Barker fought its
way out of a VC ambush, leaving 80 enemy dead.

The action began as units of the task force conducted a combat assault
into a known Viet Cong stronghold. "Shark" gunships of the 174th Aviation Company escorted the troops into the area and killed four enemy during the assault. Other choppers from the 123d Aviation Battalion killed
two enemy.

"The combat assault went like clockwork," commented LTC Frank
Barker, New Haven, Conn., the task force commander. "We had two
entire companies on the ground in less than an hour."

A company led by CPT Ernest Medina, Schofield Barracks, Hawaii,
killed 14 VC minutes after landing. They recovered two M1 rifles, a carbine, a short-wave radio and enemy documents.

Peers, *Report,* vol. 4, exhibit M-58, p. 245.

FRANK A. BARKER JR.

Combat Action Report

March 28, 1968

In a combat action report, a commanding officer reports on the planning and execution of a military operation, describes the results of the engagement, explains any difficulties encountered, and makes recommendations for subsequent expeditions. According to the report written by Lieutenant Colonel Frank Barker twelve days after the attack on My Lai, the incident had been a successful operation against a dangerous enemy. The slaughter of civilians appears nowhere in Barker's report.

TO: Commanding Officer
11th Infantry Brigade
ATTN: XIOP
APO 96217

1. *Type of Operation:* Helicopter Assault.

2. *Dates of Operation:* 160730 to 161800 Mar 68.[1]

3. *Location:* My Lai, RVN, BS 728795.

4. *Command Headquarters:* Task Force Barker, 11th Infantry Brigade.

5. *Reporting Officers:*
LTC Frank A. Barker, Jr., CO, Task Force Barker
CPT Ernest Medina, CO, Co C, 1/20 Inf
CPT Earl Nichols, CO, Co B, 4/3 Inf
CPT William Rigg, CO, Co A, 3/1 Inf

6. *Task Organizations:*
Headquarters, Task Force Barker
Company A, 3d Battalion, 1st Infantry
Company B, 4th Battalion, 3d Infantry
Company C, 1st Battalion, 20th Infantry

---

[1]7:30 A.M. to 6:00 P.M., March 16, 1968.

Peers, *Report,* vol. 4, exhibit R-2, pp. 401–05.

7. *Supporting Forces:*
Btry D, 6th Battalion, 11th Arty (105 How).
174th Avn Co (Recon Acft and gunships), timely and effective.
Co B, (Aero scout), 123d Avn Bn, timely and effective.
Coastal Surveillance Force, USN (Swift Boat) timely and effective.

8. *Intelligence:* Enemy forces in the area of operation were estimated to be one local force battalion located in the vicinity of My Lai, BS 728795 as shown in Inclosure 1. This information was based upon previous combat operations in this area, visual reconnaissance, and PW and agent reports. During the operation it was estimated that only two local force companies supported by two to three local guerrilla platoons opposed the friendly forces. The area of operation consisted of six hamlets to varying degree of ruin, each separated by rice paddies which were bounded by a series of hedge rows and tree lines. The area was also honeycombed with tunnels and bunkers. The many hedge rows offered the enemy considerable cover and concealment from the attacking friendly forces. However the clear weather permitted maximum utilization of reconnaissance aircraft and helicopter gunships to seek out and destroy enemy defensive positions.

9. *Mission:* To destroy enemy forces and fortifications in a VC base camp and to capture enemy personnel, weapons and supplies.

10. *Concept of Operation:* Task Force Barker conducts a helicopter assault on 160730 Mar 68 on a VC base camp vicinity BS 728795 with Company C, 1st Battalion, 20th Infantry landing to the west and Company B, 4th Battalion, 3d Infantry landing to the southeast of the VC base camp. Company A, 3d Battalion, 1st Infantry moves by foot to blocking positions north of the base camp prior to the helicopter assault. USN Swift Boats screen the coastal area to the east of the base camp and Company B (Aero Scout) 123d Avn Bn screens to the south to block or destroy enemy forces attempting to withdraw. See Incl 1. An artillery preparation and gunship suppressive fires are planned for both landing zones. Artillery blocking fires are planned on all paths of escape which the enemy might use. Upon landing, the two rifle companies assault enemy positions making a detailed search of all buildings, bunkers and tunnels as they move.

11. *Execution:* The order was issued on 14 March 1968. Coordination with supporting arms reconnaissance and positioning of forces was conducted on 15 Mar 68. On 160726 Mar 68 a three minute artillery prepa-

ration began on the first landing zone and at 0730 hours the first lift for Co C touched down while helicopter gunships provided suppressive fires. At 0747 hours the last lift of Co C was completed. The initial preparation resulted in 68 VC KIA's in the enemy's combat outpost positions. Co C then immediately attacked to the east receiving enemy small arms fire as they pressed forward. At 0809H a three minute artillery preparation on the second landing zone began and the first l[i]ft for Co B touched down at 0815 hours. At 0827 the last lift of Co B was completed and Co B moved to the north and east receiving only light enemy resistance initially. As Co B approached the area of the VC base camp, enemy defensive fires increased. One platoon from Co B flanked the enemy positions and engaged one enemy platoon resulting in 30 enemy KIA. Throughout the day Co B and Co C received sporadic sniper fire and encountered numerous enemy booby traps. Co A in clocking positions to the north had only light contact against small enemy elements attempting to withdraw to the north. Attempts of the enemy to escape along the beach or to the south were successfully countered by the Swift Boats and the Aero Scout Company. By 1630 hours the surviving enemy elements had broken all contact with friendly forces by infiltrating with civilians leaving the area, or by going down into the extensive tunnel systems throughout the area. At 1715 hours Co C linked-up with Co B and both units went into a perimeter defense for the night in preparation for conducting search and destroy operations the next day. With the establishment of the night defensive position at 161800 March 1968 the operation was terminated.

12. *Results:*
   a. Enemy losses:
      (1) Personnel:
         128 KIA
         11 VCS CIA
      (2) Equipment captured:
         1 M-1 rifle
         2 M-1 carbines
         10 Chicom hand grenades
         8 US M-26 hand grenades
       410 rounds small arms ammo
         4 US steel helmets with liners
         5 US canteens with covers
         7 US pistol belts
         9 sets US web equipment

2 short wave transistor radios
3 boxes of medical supplies
  (3) Equipment and facilities destroyed:
     16 booby traps
      1 large tunnel complex
     14 small tunnel complexes
      8 bunkers
    numerous sets of web equipment
  b. Friendly losses:
     2 US KHA
    11 US WHA

13. *Administrative Matters:*

a. Supply. Units moved with basic loads of ammunition and three C-ration meals per man. Resupply was planned and effected by helicopter. No problem existed in resupply.

b. Maintenance. No problems encountered.

c. Medical treatment and evacuation. All casualties requiring evacuation were removed from the area by helicopters including wounded VC and some of their civilian supporters. All other casualties were treated by company aidmen.

d. Transportation. Helicopters were the primary means of transportation. No problems encountered.

e. Communications. No problems encountered.

14. *Special Equipment and Techniques:*

a. Aero Scout Company. This unit was used effectively as a reconnaissance and supporting force along the southern portion of the area of operation.

b. US Navy Swift Boats. Effective use of these craft was made to provide surveillance of the beach area and to detect enemy personnel attempting to escape in boats or along the beach.

15. *Commander Analysis:* This operation was well planned, well executed and successful. Friendly casualties were light and the enemy suffered heavily. On this operation the civilian population supporting the VC in the area numbered approximately 200. This created a problem in population control and medical care of those civilians caught in fires of the opposing forces. However, the infantry unit on the ground and helicopters were able to assist civilians in leaving the area and in caring for and/or evacuating the wounded.

16. *Recommendations:* Operations conducted in an area where large numbers of refugees might be generated should provide for civil affairs, psyops, medical, intelligence and police teams to be brought to the area as early as practicable after the arrival of combat troops. This would facilitate population control and medical care, and would permit the sorting out of VC which have mingled among the population for cover. The presence of these teams would free infantry personnel for combat operations.

FRANK A. BARKER, JR.
Lt Colonel, Infantry
Commanding

3

WILLIAM WESTMORELAND

Testimony before Congressional Committee
1970

Barker's combat action report, though completely fraudulent, made its way up the chain of command, at each level earning the requisite praise for a "job well done." General William Westmoreland, as commander of the Military Assistance Command, Vietnam (MACV), from June 1964 to July 1968, sat at the top of the command structure. During his testimony before the House Armed Services Committee in 1970, Westmoreland said he vaguely remembered the first reports he received concerning the assault on My Lai on March 16, 1968.

As to the reports of the 11th Brigade's operations around Son My in March 1968, the picture my headquarters in Saigon received was favorable. It appeared that the operation—with 128 Viet Cong killed and 3

House Committee on Armed Services, Armed Services Investigating Subcommittee, *Investigation of the My Lai Incident: Hearings of the Armed Services Investigating Subcommittee of the Committee on Armed Services,* 91st Cong., 2nd sess., 1970, 837.

weapons captured at the cost of only 2 U.S. soldiers killed—was a tactical success. In fact, from the operational reports we received, it appeared so successful that I responded with a congratulatory message. Such messages were habitually sent in those instances of conspicuous tactical success. During the first 3 months of 1968, for example, 47 messages of that type were sent to units in the field in my name.

2

The Official Rules
of Modern Warfare

In 1863, at the height of the Civil War and at the behest of the War Department, the distinguished political scientist Francis Lieber undertook a codification of existing international law on the rules of land warfare. The War Department published Lieber's work as *A Code for the Government of Armies in the Field* and instructed Union soldiers to conduct themselves according to its conclusions. The code's guiding principle was as follows: "Men who take up arms against one another in public war do not cease on this account to be moral beings, responsible to one another and to God."

Since that time, a body of international law has evolved concerning the treatment of prisoners of war and the treatment of noncombatants during war. Those laws, which the United States has ratified, establish the legal behavior of soldiers engaged in combat operations and define the nature of war crimes. In 1907, the Hague Conference formulated an international code, the Hague Convention on Land Warfare, which calls for the humane treatment of prisoners of war and prohibits the wanton killing of noncombatants and unnecessary destruction of their property.

After World War II, when the Allied powers held the Nuremberg trials to prosecute many former Nazis for war crimes, many defendants claimed innocence because their military superiors had ordered them to commit the crimes. To prevent such a criminal defense in the future, the Nuremberg tribunal proclaimed what has become known as the "Nuremberg principle": that "the fact that a person acted pursuant to order of his Government or of a superior does not relieve him from responsibility under international law, provided a moral choice was in fact possible to him."

Three years later, leaders from around the world met in Geneva, Switzerland, and negotiated the Geneva Conventions on the Laws of War. Among other items, the Geneva conventions require the humane treatment of all civilians and prisoners of war; prohibit murder, mutilation, torture, the taking of hostages, and offenses against personal dignity; require medical attention for the sick and wounded; and outlaw the passing of sentences and the carrying out of executions without due process of law. As William Calley and his men entered My Lai on March 16, 1968, they were operating under the Geneva conventions.

4

WILLIAM WESTMORELAND

Testimony before Congressional Committee
1970

In his testimony before the House Armed Services Investigating Committee in 1970, General William Westmoreland, former MACV commander, explained the program MACV employed to train all members of the armed forces in the proper treatment of noncombatants, the rules of engagement, and the nature of war crimes. He argued that the reality of the war in Vietnam — a guerrilla conflict in which civilians often assisted the enemy or actually served in Vietcong units — made the goal of providing humane treatment of all noncombatants especially challenging. Nevertheless, Westmoreland believed that the U.S. army had indeed fulfilled its mission of educating American soldiers.

Among the policies I established as the commander in Vietnam were detailed rules of engagement—particularly concerning the use of our firepower—as well as instructions on minimizing noncombatant casualties and procedures for the reporting and investigation of war crimes. I considered it extremely important that our policies on these subjects be carefully spelled out in Vietnam, because of the peculiarities of the conflict there. For one thing, enemy forces were frequently intermingled with the civilian populace. For another, our forces were "guests" in a foreign country, and their conduct therefore had to be exemplary.

House Committee on Armed Services, Armed Services Investigating Subcommittee, *Investigation of the My Lai Incident: Hearings of the Armed Services Investigating Subcommittee of the Committee on Armed Services,* 91st Cong., 2nd sess., 1970, 834–36.

1. Geneva Convention Training

All soldiers were required to receive 1 hour of training in the Geneva and Hague Conventions during basic training. In addition, Army regulations require that qualified legal officers conduct refresher training in this subject once each year. Every replacement arriving in Vietnam was also given several wallet-size cards containing instructions pertinent to this and related matters. Two of these concern the treatment of noncombatants and prisoners of war. They are entitled "Nine Rules" and "The Enemy in Your Hands." Copies of these cards have been provided to your subcommittee. These cards stressed humanitarian treatment and respect for the Vietnamese people, and stipulated that each individual would comply with the Geneva Convention. Additionally, commanders down to battalion level received a card entitled "Guidance for Commanders in Vietnam" which, among other points, emphasized the commander's responsibility for the conduct of his subordinates.

2. Rules of Engagement

These rules are based on guidance provided by the Joint Chiefs of Staff. Because of the constant turnover of personnel in Vietnam, I established a policy in 1966 of frequent review, revision, and republication of the rules of engagement. This was to insure maximum visibility to all U.S. personnel during their tour of duty, and was done at least once each year. These rules provided specific guidance for the conduct of combat operations—particularly the control of firepower—and directed that all possible measures be taken to reduce the risk to the lives and property of friendly forces and civilians.

3. Treatment of Noncombatants and Prisoners of War

A series of directives were published which sought to minimize casualties among noncombatants, to protect the property of Vietnamese citizens, and to preserve the rights of those persons captured by us. I wanted commanders at all levels to become involved in these matters, since so much of our success in Vietnam depended on winning and retaining the respect of the Vietnamese people. To that end, I required that commanders repeatedly emphasize to their troops both the short-range and long-range importance of minimizing civilian casualties. I also required that each combat operation be preceded by a briefing which outlined the procedures for safeguarding noncombatants and their property.

4. War Crimes

My directives covering war crimes, in addition to defining the term, cited examples of incidents which could be considered "grave breaches" of the Geneva Convention. The directives required that anyone having knowledge of an incident or act thought to be a war crime report it to his commander. The commander, in turn, was required to report this matter to my staff at MACV headquarters. Procedures were established for the investigation of all such incidents, under the direction of my staff judge advocate.

5. Serious Incidents

As I indicated earlier, the nature of the Vietnam war created special hazards for noncombatants. The infiltration of the nonuniformed enemy into the local populace made identification difficult, and increased the likelihood of injury to innocent people. Early in the conflict these factors, and many others associated with this unique war, caused me great concern. I wanted to know about each incident as it occurred, so that we could benefit from our experience and correct our mistakes. As a result, we published a directive in September 1966 which required that my headquarters be notified of any incident involving major property loss; death, injury, or mistreatment of noncombatants; or the killing, wounding, or mistreatment of friendly personnel by United States, Vietnamese, or free world forces. The directive was designed to cover incidents not specifically mentioned in other MACV directives. . . .

My basic attitude on noncombatant casualties—and that of my command—were well expressed in a statement I made to the press in August 1966. A copy of that statement has been furnished to your subcommittee. In it, I stated that, "one mishap—one innocent civilian killed, one civilian wounded, or one dwelling needlessly destroyed, is too many. I emphasized to the press that, "we are sensitive to these incidents and want no more of them," and that we were attacking the problem aggressively. To make sure that my message to the press also got to the troops, at a meeting of my commanders a few days later I gave each of them a copy of the rules of engagement, reiterated the statement I had made to the press, and directed that my commanders insure that their subordinates were thoroughly familiar with the appropriate regulations.

Periodically at meetings with my subordinate commanders I would discuss this matter to insure that new commanders and newly arrived troops were aware of the importance that I attached to troops' conduct and avoidance of civilian casualties. For example, on December 3, 1967, I discussed these matters in a long meeting and in rather strong terms with my senior commanders who had assembled at Nha Trang. I again directed that my commanders take a personal interest in the attitude and conduct of our troops in their dealings with the Vietnamese and the importance of minimizing casualties among the civilian populace. I am leaving with your subcommittee staff a summary of that discussion, to add to the background papers you already have.

Later, during the Tet offensive in early 1968, photos and articles appeared in the press in which newsmen reported alleged mistreatment of civilians and prisoners of war. On February 21, 1968, I cited such reports in a message to all of my commanders, stating in clear terms that such actions would not be condoned. This message was also dispatched to U.S. advisory personnel directing they make every effort to influence their Vietnamese counterparts to observe the rules which we had adopted.

On February 21, 1968, the day I sent the message to the field, I sent a copy of this message to General Vien, chief of the joint general staff of the Vietnamese Armed Forces, pointing out my concern over these matters and encouraging him to take similar action within Vietnamese channels.

MACV also made extensive use of the command information media to get word of its policies to troops in the field. In place of "commercials," the Armed Forces radio and television network used spot announcements. Frequently these concerned methods of handling prisoners of war, the importance of proper individual conduct, and our relationship with the Vietnamese. Similar items were included in the MACV newspaper, The Observer; which was distributed to all units in Vietnam.

5

Wallet Cards

Throughout the war, the Military Assistance Command, Vietnam (MACV) tried to distribute information on the treatment of civilians to each American soldier serving in Vietnam. Officers and enlisted personnel received a wallet-sized card entitled "Nine Rules for Personnel of US Military Assis-

Peers, *Report,* vol. 4, exhibits M-2, M-3, pp. 9, 11.

tance Command, Vietnam." Another handout—"The Enemy in Your Hands"—was also distributed widely among the military during the war.

NINE RULES

For Personnel of US Military Assistance Command, Vietnam

The Vietnamese have paid a heavy price in suffering for their long fight against the communists. We military men are in Vietnam now because their government has asked us to help its soldiers and people in winning their struggle. The Viet Cong will attempt to turn the Vietnamese people against you. You can defeat them at every turn by the strength, understanding, and generosity you display with the people. Here are nine simple rules:

Nine Rules

1. Remember we are guests here: We make no demands and seek no special treatment.
2. Join with the people! Understand their life, use phrases from their language and honor their customs and laws.
3. Treat women with politeness and respect.
4. Make personal friends among the soldiers and common people.
5. Always give the Vietnamese the right of way.
6. Be alert to security and ready to react with your military skill.
7. Don't attract attention by loud, rude or unusual behavior.
8. Avoid separating yourself from the people by a display of wealth or privilege.
9. Above all else you are members of the US Military Forces on a difficult mission, responsible for all your official and personal actions. Reflect honor upon yourself and the United States of America.

DISTRIBUTION—1 to each member of the United States Armed Forces in Vietnam

THE ENEMY IN YOUR HANDS

1. *Handle him firmly, promptly, but humanely.* The captive in your hands must be *disarmed, searched,* secured and watched. But he must also be treated at all times as a human being. He must not be tortured, killed,

mutilated, or degraded, even if he refuses to talk. If the captive is a woman, treat her with all respect due her sex.

2. *Take the captive quickly to security.* As soon as possible evacuate the captive to a place of safety and interrogation designated by your commander. Military documents taken from the captive are also sent to the interrogators, but the captive will keep his personal equipment except weapons.

3. *Mistreatment of any captive is a criminal offense. Every soldier is personally responsible for the enemy in his hands.* It is both dishonorable and foolish to mistreat a captive. It is also a punishable offense. Not even a beaten enemy will surrender if he knows his captors will torture or kill him. He will resist and make his capture more costly. Fair treatment of captives encourages the enemy to surrender.

4. *Treat the sick and wounded captive as best you can.* The captive saved may be an intelligence source. In any case he is a human being and must be treated like one. The soldier who ignores the sick and wounded degrades his uniform.

5. *All persons in your hands, whether suspects, civilians, or combat captives, must be protected against violence, insults, curiosity, and reprisals of any kind.* Leave punishment to the courts and judges. The soldier shows his strength by his fairness, firmness, and humanity to the persons in his hands.

Key Phrases

ENGLISH	VIETNAMESE
Halt	Đứng lại
Lay down your gun	Buông súng xuống
Put up your hands	Đưa tay lên
Keep your hands on your head	Đưa tay lên đầu
I will search you	Tôi khám ông
Do not talk	Đừng nói chuyện
Walk there	Lại đằng kia
Turn Right	Xoay bên phải
Turn Left	Xoay bên trái

"The courage and skill of our men in battle will be matched by their magnanimity when the battle ends. And all American military action in Vietnam will stop as soon as aggression by others is stopped."

21 August 1965   Lyndon B. Johnson

The Enemy in Your Hands

As a member of the US military forces, you will comply with the Geneva Prisoner of War Conventions of 1949 to which your country adheres. Under these conventions:

YOU CAN AND WILL

Disarm your prisoner
Immediately search him thoroughly
Require him to be silent
Segregate him from other prisoners
Guard him carefully
Take him to the place designated by your commander

YOU CANNOT AND MUST NOT

Mistreat your prisoner
Humiliate or degrade him
Take any of his personal effects which do not have significant
    military value
Refuse him medical treatment if required and available

ALWAYS TREAT YOUR PRISONER HUMANELY

TRAINING IN THE RULES
OF LAND WARFARE

By the time of the Vietnam War, international law carefully delineated the proper behavior of soldiers during land warfare. In 1968, official U.S. army regulations called for specific training of all soldiers in the rules of land warfare and in the appropriate treatment of noncombatants. During his commission's investigation of the My Lai incident, Lieutenant General William Peers tried to assess the quality of that training. Most of the troops who were interviewed as part of the investigation were asked specifically if they could remember being trained in the laws of land warfare. Since months and even years had passed since their training, only a few of them had even the vaguest recollection of being instructed in the nature of war crimes. They were much more likely to remember being told to obey—absolutely and without reservation—the orders of military superiors.

HERBERT L. CARTER

Testimony to Peers Commission
1970

Herbert Carter was a "tunnel rat" with Charlie Company. Along with many other soldiers, Carter could barely remember any training in the rules of warfare, but he did remember that the instructor treated the lesson in a joking manner, as if the entire idea of rules in war was absurd.

Q: Do you recall getting any instructions on how to handle prisoners of war during this time?

A: We had a little instruction on that.

Q: Do you recall what they told you?

A: They told us that if we get a prisoner to hold them until someone, intelligence, was actually supposed to interrogate them. The instructor sort of laughed about this.

Q: Why did they, or he, laugh about this? Do you know?

A: It was just the way they said it, like you do what you want to do with them actually.

Peers, *Report*, vol. 2, bk. 24, p. 3.

PEERS COMMISSION

Training of 11th Brigade
1970

After listening to the My Lai participants describe the training they had received in the rules of land warfare and the treatment of noncombatants, Peers investigators concluded that the Americal Division and its 11th Brigade had been so hastily assembled in 1967 that proper training had not taken place.

Lack of Emphasis in Training

Early in the Inquiry, there was a suspicion that the manner in which the 11th Brigade was activated, trained, prepared for overseas movement, and deployed to Vietnam might have had some impact upon the events of Son My. Investigation revealed that this was the case to a limited extent.

11th Brigade elements underwent an accelerated training program, received a substantial input of replacement personnel shortly before deploying, and eventually deployed earlier than originally had been scheduled. Shortly after arriving in Vietnam, planned makeup training was effected by another infusion of replacements (to overcome a projected rotation "hump") and by early commitment of brigade elements to active combat operations.

As a net result of these actions, the evidence indicates that, at best, the soldiers of TF Barker had received only marginal training in several key areas prior to the Son My operation. These areas were (1) provisions of the Geneva Conventions, (2) handling and safeguarding of noncombatants, and (3) rules of engagement.

The problem of training and instruction having to do with identification of and response to "illegal" orders is addressed elsewhere in this report. The evidence indicates that training deficiencies in this area, together with deficiencies in those training areas described above, played a significant part in the Son My operation.

Peers, *Report,* vol. 1, pp. 8–13.

3

Experience in War
December 1967 to March 14, 1968

The 23rd Infantry Division, also known as the Americal Division, had been deactivated in April 1956 after several years of service in the Panama Canal Zone. General William Westmoreland reassembled and reactivated the unit on September 25, 1967. He assigned Americal troops to I Corps in Vietnam, where the unit was to work cooperatively with the 1st Marine Division. Charlie Company, part of Americal's 11th Infantry Brigade, was deployed to Quang Ngai province in I Corps in December 1967. For the next three months, the troops participated in search and destroy missions, trying to locate the notorious Vietcong 48th Local Force Battalion, which operated regularly in the region.

It was an extremely difficult time for the soldiers. As you analyze the testimony in the following documents, look for frustration among the troops over their inability to locate and engage the Vietcong, even while they sustained serious casualties from mines and booby traps. You should also be able to detect a deep sense of resentment accumulating toward South Vietnamese civilians. The members of Charlie Company were convinced that villagers knew where the Vietcong had placed the mines and booby traps but refused to inform or even warn the Americans. Some members of Charlie Company eventually came to see all Vietnamese — North Vietnamese, South Vietnamese, and the Vietcong — as the enemy.

8

PEERS COMMISSION

Military Situation in Quang Ngai Province

1970

When Charlie Company deployed to Quang Ngai province in December 1967, the soldiers entered a particularly dangerous region of South Vietnam, an area where enemy military forces were unusually strong and where civilian sympathies tended to support the North Vietnamese and the Vietcong rather than the United States or the government of South Vietnam. The Peers Commission developed the following assessment of the military situation that existed in Quang Ngai province at the time of the My Lai attack.

Enemy Situation in March 1968

As a basis for evaluating the enemy situation in Quang Ngai Province, it is noted that in March 1968, enemy strength throughout South Vietnam was estimated to be approximately 263,200 men. Of this total, about 55,900 were Viet Cong (VC) combat forces, 87,400 were North Vietnamese Army (NVA) combat troops, and 69,100 were guerrillas, with the remaining 50,800 comprising administrative personnel. . . .

Considering only Quang Ngai Province . . . , enemy strength ranged between 10,000 and 20,000 men during the 4 years preceding the Son My incident. In early 1968 enemy strength was estimated to be between 10,000 and 14,000 men of which 2,000–4,000 were regular forces, 3,000–5,000 were guerrillas, and 5,000 were assigned to administrative units.

A number of VC and NVA regiments operated in Quang Ngai Province from 1964 to 1966. However, four local force battalions and eleven companies of VC were the forces primarily responsible for harassing the area under government control. The 48th Local Force (LF) Battalion became the principal enemy force in Son Tinh District, although it also operated in the Batangan area to the north as well as to the south of the Song Tra Khuc. Members of the 48th LF Battalion reportedly lived with the local villagers in order to conceal their presence, often working as farmers during the day and fighting as guerrillas at night. . . .

Peers, *Report*, vol. 1, pp. 3-1–3-6.

... In the *Tet* operation the 48th LF Battalion overran the Regional Force/Popular Force Training Center near Son Tinh and held it briefly until driven out by counterattacking 2d ARVN Division forces. In the ensuing fight the 48th LF Battalion reportedly suffered about 150 casualties, including the battalion commander and two company commanders, and a third company commander captured.

With the failure of the assault of Quang Ngai City and other province towns [during the Tet Offensive], VC units filtered back to their home areas, mostly to the south and west. Because of its heavy losses during *Tet,* elements of the 48th LF Battalion withdrew to the mountains in western Quang Ngai to reorganize and refit, while other elements of the battalion returned to their habitual area of operation on the Batangan Peninsula. By late February, the 48th LF Battalion headquarters had reportedly returned to the peninsula, but the unit remained out of contact during the first part of March, apparently to continue recuperating from the *Tet* setback. At the time of the Son My incident, the 48th LF Battalion had an estimated strength of 200–250 and was the only major enemy unit with elements in the Son My area. However, there were two additional local force companies in the district which on occasion joined the 48th LF Battalion in carrying out specific operations. Overall guerrilla strength in Son Tinh district was reported to be about 700 strong.

Son My Village

Son My Village is located approximately 9 kilometers northeast of Quang Ngai City and fronts on the South China Sea. In March 1968, the village was composed of four hamlets, Tu Cung, My Lai, My Khe, and Co Luy, each of which contained several subhamlets.... Most of the residents of Son My either farm the rich alluvial soil along the rivers and streams or engage in offshore fishing operations.

The People of Quang Ngai Province

Historically, the people of Quang Ngai Province have a long record of supporting rebellion. In the 19th century they had been a focal point of resistance to French control of Indochina. Later, in the 1930's, they had fomented peasant revolts against Vietnamese supporting the French. After World War II when the French sought to reestablish themselves in Indochina, Quang Ngai became a Viet Minh stronghold and by 1948 Ho Chi Minh considered it free from French rule. Duc Pho, in southern Quang Ngai, became one of the largest rest and recreation areas for the

Viet Minh forces until the country was divided by the Geneva Accords in 1954.

Although most of the Viet Minh departed for the north after the settlement of the Geneva Accords, some remained behind and their influence was particularly strong in the rural areas. By the 1960's, a whole generation of young people had grown up under the control of the Viet Minh and the later National Liberation Front....

... In the eyes of the Government of Vietnam (GVN) the people who continued to live in the Son My area were considered generally to be either VC or VC sympathizers.

Enemy Tactics and Techniques

As previously discussed, the enemy forces which operated in Quang Ngai Province and Son Tinh District included guerrillas, local and main force units and, at times, NVA units. These forces were highly skilled in hit-and-run guerrilla tactics and had the ability to survive in a counterinsurgency environment....

Regardless of the type unit, the tactics employed by the Communist forces recognized their own shortcomings and were designed to exploit the weaknesses of the US, ARVN and other Free World Military Assistance Forces. Lacking the strength and firepower to survive an extended major battle, they relied primarily on operations which permitted them to mass, attack, and withdraw before US or GVN/ARVN forces could react. Their operations at every level were characterized by methodical planning, detailed rehearsals, and violent execution.

Prior to undertaking an operation, the VC/NVA normally would obtain very detailed information regarding their potential targets including the location of fighting positions, key installations, and the identification of security weaknesses. Using this information, which might require weeks or months to develop, they would then prepare a detailed step-by-step plan for the operation. The plan would then be rehearsed until every man in the force was thoroughly familiar with details of the target area and the functions he was to perform.

The VC had the choice of the time they wanted to fight and were willing to delay execution of an operation for as long as necessary in order to improve their chances of success. Once the decision was made to attack, the unit was moved, using clandestine techniques, to the target area. In doing this, the VC would often attempt to infiltrate demolitionist, sapper type personnel into the area to destroy key installations, and artillery and automatic weapons positions. Their final attack normally was executed only at a predetermined time or after the presence of their infiltrators had been detected. As an alternate type of attack they sometimes employed mortars, rockets, and recoilless rifles in stand-off attacks against population centers and military installations to prepare or soften the target for attack. These same basic procedures were generally followed in every type of operation, operations characterized by stealth, surprise, and shock action.

Typical operations conducted at the local force level included the ambushing of small convoys, attacking of village and district offices or security outposts, the assassination or kidnapping of local Vietnamese officials and other acts designed to illustrate their control of the area in which they operated. The main force and NVA units assisted the local force units but primarily conducted large-scale operations against US and ARVN forces and installations.

The VC made extensive use of mines and boobytraps, especially at the hamlet and village level. In addition to the men in their combat units, children, women, and old men were used to construct homemade boobytraps and mines which they normally emplaced at night under the cover of darkness. The mines and boobytraps were used in a wide variety of ways. Some of them were employed as weapons of terror against the population, such as mines planted under or along well used roadways to blow-up buses and other vehicles; demolition devices installed in theaters and other crowded areas; or a simple grenade thrown into a group of people. In another tactic, they used them as defensive weapons to cover roads, paths, and other avenues of approach to and within their controlled areas. Some such areas were literally infested with VC mines and boobytraps and had the effect of slowing and restricting friendly offensive operations. It was this latter type of employment which tended to create hatred and frustration against the unseen enemy.

The operations of all VC/NVA forces in a particular area were closely controlled and coordinated with the local VC infrastructure's political and administrative apparatus in the attempt to achieve their objective of total domination of the people. The Communist[s] recognized but few restraints in their operations and were often ruthless in conducting them. All operations were planned and executed keeping in mind the ultimate goal of seizing control of the government of South Vietnam and the people.

9

MICHAEL BERNHARDT

Testimony to Peers Commission
1970

It is common in warfare for soldiers to dehumanize the enemy, and Vietnam was no exception. American soldiers often felt alienated from the South Vietnamese, whom they were ostensibly protecting from communism. The difficulty of distinguishing between enemy troops and civilian noncombatants, racist attitudes toward Asians on the part of many American soldiers, and the frequency with which GIs were wounded or killed by booby traps, mines, and snipers all created an atmosphere of hostility between American soldiers and South Vietnamese civilians. Michael Bernhardt, a member of Charlie Company who spoke some Vietnamese, explained that attitude in his testimony to the Peers Commission.

Bernhardt's testimony also raises the most troubling and controversial aspect of My Lai—whether the massacre was a rare, isolated incident or just one particularly egregious example of a large, sustained pattern of American war crimes against South Vietnamese civilians. Historians and Vietnam veterans still debate that issue today.

Q: Why were you sent away on detail when the investigating officer was coming?

A: The investigating officer wasn't there to talk to everyone. We were spread out along the area of the perimeter. But I still think I might have been sent away because I might have said something that would have damaged the company.

Q: You mean . . . that you didn't approve of what was going on, or what had been going on [the My Lai attack]?

A: That's what I mean, yes, sir.

Q: And this was known to other people in the company?

A: Yes, sir.

Q: Did you explain why?

A: I believe I did, sir, but it wasn't getting through to too many people.

Q: You believe you did what?

Peers, *Report,* vol. 2, bk. 25, pp. 20–22, 38–39.

A: Tried to explain why I didn't think it was right. It didn't have strategic value to it at all. I believe that there is an effort on the part of some people in the higher quarters of the military, both allied and American, who want to see the war dragged on. One of the ways that they do this is by cultivating this attitude that someone very aptly called the "dink complex," and that these things that happened as a result of the "dink complex" damage us. The idea I believe in that war is not to kill off the enemy. It would just be too hard to do that. You have got to make them want to stop fighting or else eliminate his means of fighting. When you go out and do something like this, I believe what you are doing is breeding more Viet Cong.

Q: You mean by killing the civilians at My Lai (4)?

A: Yes, sir.

Q: Go ahead.

A: I believe that this breeds Viet Cong and this isn't helping us at all. It is more hurting us. That's also why—and I want to make it clear here—that's why I made these public statements and so on. I wasn't trying to drag anybody down. I think that already now an attitude has changed by men who either are serving or will serve in Vietnam—that they won't get away with this all the time. It was an attitude that was prevalent in my company, and I wanted to see it reduced altogether.

Q: When you speak of a "dink complex" were you referring to how American soldiers look upon the Vietnamese people?

A: Yes, sir.

Q: Could you explain that a little more. I think I know what you mean, but I'm not sure.

A: All right, sir. From the way I look at it, I believe that it is mostly the linguistics bit. Now a person loses a certain aspect for being valued as a human being if you cannot understand him—rather if he doesn't have the means of communication with somebody else. A lot of the men—to their way of thinking, since the Vietnamese were speaking something that we could not understand, felt that they weren't communicating with anyone. It is just a sort of psychological way to look at it. What they thought were these people were a whole lot less than human. They knew, or they at least heard, of their value of human life. I think we're stuck with our values. Also they could get away with just about anything that they wanted to get away with. There is a lot of frustration that is among the men over there, and these frustrations cannot be directed at those responsible for creating them, and so, they're directed at what they can be directed at. In other words, making sort of a whipping boy out of the South Vietnamese population.

Q: In other words, helpless people—they just take it out on them?

A: Yes, sir, it sounds—it's illogical, of course, but we're not going to look for logic in a large number of men that would do this.

Q: Is this attitude related to the practice of referring to the Vietnamese as "dinks," "slopes," and "gooks," and this sort of thing?

A: Yes, sir, right, something like that. These are small manifestations of it. The larger manifestations is this and also the fact that there are a lot of—I said to the press that I thought it was an isolated incident. I didn't think it was an isolated incident, but I didn't think it would be wise to tell the press that I thought it happened all over Vietnam, because, first of all, I didn't know and, second of all, it wouldn't help any to say that it did.

Q: You feel that this is not isolated, but you don't actually have evidence of other—

A: (Interposing) Right. All I have is what information I've gathered by talking to other men about this particular incident and also not relating to this incident, but just talking about Vietnam in general. . . .

Q: . . . Did you understand, or did you comprehend the difference between civilians and noncombatants, detainees?

A: I understood myself, sir. I just don't—I don't know about the other men, but I understood myself. I didn't see any reason to ask questions, because it seemed to me that I knew about as much as anybody I could ask. Maybe it sounds presumptuous, but I didn't think that Lieutenant Calley or Captain Medina could have provided me with a better answer than I had myself. . . .

Q: . . . What do you mean by this, you tried something similar to this before or what?

A: No, sir. But, once when we started operating and when we first came into contact with Vietnamese civilians, or whatever they were, apparently civilians, there were several old women with "chogie sticks"— that is, the long stick that they balance on their shoulder with two baskets, one on either end—in front of me. I ordered them to stop, that is "dung lai," which means stop, I think. I never got the right reaction out of "dung lai," and they kept on going only a little bit faster. I fired some shots over their heads. After that time Lieutenant Calley said to me that "the old man" says—which could mean any "old man," the company commander, the battalion commander—said that if they don't stop when you say "dung lai," you shoot them. So I didn't think of this as really being the best way to handle it. There could be any reason why they don't stop when I say "dung lai." Not knowing for sure what "dung lai" means myself, I might be telling them to get lost. It's a tonal language. You have to know the music, not only the words, so I might be saying it wrong. I just couldn't see doing that. I was told that they might have hand grenades in the baskets and I thought they might also have fish, since they were fish baskets, and I didn't really trust the judgment of the people that I had in command of me.

10

WILLIAM L. CALLEY

Combat Experiences before My Lai
1970

At his court-martial trial, William Calley was asked about casualties inflicted on his men during combat operations prior to My Lai and how those events affected his attitude.

Q: Everytime that the company would go, at least a company-sized unit, to try to get in that area and stay in there, they encountered hostile fire, enemy fire, suffered casualties, and were driven out?

A: Yes, sir. [Calley was asked about an incident that occurred when he was returning to his company from in-country R and R. As he was waiting for a helicopter to take him to his men, he helped unload a chopper filled with casualties caused by a mine field.]

Q: What did you see and what did you do in connection with that helicopter when it landed back there and before you boarded up to go to meet your company?

A: The chopper was filled with gear, rifles, rucksacks. I think the most — the thing that really hit me hard was the heavy boots. There must have been six boots there with the feet still in them, brains all over the place, and everything was saturated with blood, rifles blown in half. I believe there was one arm on it and a piece of a man's face, half of a man's face was on the chopper with the gear.

William Calley Court-Martial Transcripts, National Archives Complex, College Park, Maryland, pp. 3785–86.

Q: Did you later subsequently learn that those members that were ema-
ciated in that manner were members of your company or your platoon?

A: I knew at that time they were.

Q: What was your feeling when you saw what you did see in the chopper
and what you found out about your organization being involved in that
kind of an operation?

A: I don't know if I can describe the feelings.

Q: At least try.

A: It's anger, hate, fear, generally sick to your stomach, hurt.

Q: Did it have any impact on your beliefs, your ideas or what you might
like to do in connection with somehow or other on into combat and
accomplishing your mission? Am I making that too complicated for
you?

A: I believe so.

Q: I'm trying to find out if it had any impact on your future actions as you
were going to have to go in and if you did go in and reach the enemy
on other occasions and if so, what was the impact?

A: I'm not really sure of what my actual feelings were at that time. I can't
sit down and say I made any formal conclusions of what I would do
when I met the enemy. I think there is an—that instilled a deeper
sense of hatred for the enemy. I don't think I ever made up my mind
or came to any conclusion as to what I'd do to the enemy.

Q: All right. Now did you have any remorse or grief or anything?

A: Yes, sir, I did.

Q: What was that?

A: The remorse for losing my men in the mine field. The remorse that
those men ever had to go to Vietnam, the remorse of being in that situation where you are completely helpless. I think I felt mainly remorse
because I wasn't there, although there was nothing I could do. There
was a psychological factor of just not being there when everything is
happening.

Q: Did you feel sorry that you weren't there with your troops?

A: Yes, sir.

11

JAY A. ROBERTS

Testimony to Peers Commission
1970

Peers Commission investigators asked Jay Roberts, an army journalist who accompanied photographer Ronald Haeberle on the My Lai operation, to explain why the massacre had occurred. Roberts was a veteran journalist, and My Lai was not his first assignment. He had witnessed My Lai and then had written a completely sanitized version of the massacre (see p. 27).

Roberts's testimony raises troubling questions. Was such behavior typical of army journalists? Had he seen other massacres and treated them just as cavalierly? Was My Lai the exception or the rule during the Vietnam War?

Q: I would like to ask you two questions, Mr. Roberts, that are very germane to this investigation that we are conducting. The first question is, why did this thing happen?

A: Well, that's a very difficult question and I have pondered since this thing was brought out in the press. I think basically it was a reaction to frustration by these people that were in this operation. I don't think that they had orders to go in there and shoot women and children. I think they had orders to go in there and clean out this VC nest, and I think that they expected to have a lot of resistance and to be really in a heavy fire fight. They had been in the same area two other times in the past 2 weeks, and they really had been in a lot of trouble. I think the press brought out the fact that they had a little memorial service for one of their people who was quite well liked. It was rather an emotional little service and it probably helped to heighten their hatred for this area and the VC in this area. Plus the fact that any GI in Vietnam is in a frustrated situation. He doesn't know who to be friends with. Children coddle up to jeeps and drop hand grenades in them. You can't trust a child because anybody in Vietnam—because you don't know who your friends are. I think that these things were working on these people. The situation was right and they went in there to clean out this VC nest and some of the individuals among the group got carried

---

away. In every large group you find some hostile people and some don't-care-type people, and I think the hostile people, the I-don't-care people did what was done there. It was just a bad reaction to these instances and the situation in Vietnam.

Q: You also indicated in your earlier discussion that these pep talks that they received, did that also seem to add fuel to fire, so to speak?

A: I think generally that that was a considerable factor. Also, the fact that they really had made no headway in the two previous engagements and the fact that they had taken quite a few casualties. The fact that this time they were going to be ready and go back in there; nothing was going to stop them, which is the type of thing that I would think would be a good way to brief guys when they're about to go into a heavy fire fight, where they would expect to have their friends dropping all around them. "Don't stop for anything. We're going to take this hill," that type of thing, and that probably was the type of pep talk that I referred to earlier, although I don't know.

Q: But the men were worked up to quite a high pitch?

A: From what I understand, from what I had heard, they weren't going to be easily put down.

4

The Briefings
March 15, 1968

After completing their investigation of the My Lai incident, members of the Peers Commission tried to provide information on how the atrocity could have been committed by a U.S. army platoon. They identified a number of factors—the difficulty in engaging the enemy, the nature of guerrilla warfare, and the problem of distinguishing between enemy soldiers and civilians—that contributed to the rage, frustration, and fear felt by many members of Charlie Company.

Army investigators were particularly concerned about the briefings that took place in Charlie Company on March 15, 1968, the day before the assault on My Lai. The Geneva conventions and U.S. army regulations require soldiers to disobey orders to commit war crimes and to report such orders to appropriate authorities in a timely manner. The investigators wanted to determine if Lieutenant Colonel Frank Barker had intended to bring about the complete destruction of My Lai and all of its inhabitants and whether such an order had been communicated down the chain of command; or if soldiers at the platoon and squad levels had somehow misinterpreted their mission and, if so, how such confusion had occurred; or whether clandestine enemy activity had triggered the massacre.

In any case, the pre-operation briefings were critical in understanding what happened at My Lai. Did Barker order Captain Ernest Medina to destroy My Lai? Did Medina tell his platoon leaders, including Lieutenant William Calley, to do so? Did Calley really order his troops to annihilate the hamlet and its residents? And, if so, how morally guilty were the troops for carrying out the orders? As you can see from the testimony in this chapter, exactly what took place at the briefings is all but impossible to reconstruct.

12

PEERS COMMISSION

Pre-Operation Briefings
1970

In their inquiry into My Lai, Peers investigators were especially concerned with pre-operation planning. They hoped to learn exactly what Lieutenant Colonel Frank Barker had hoped to achieve and how he had communicated those objectives to his company commanders. Lieutenant General William Peers also wanted to learn what specific instructions Captain Ernest Medina had given to the men in his company. The following document expresses the commission's conclusions about Task Force Barker's plans and orders.

General

In reviewing the events which led up to the Son My operation of 16 March 1968 and the military situation that existed in the area at that time, certain facts and factors have been identified as having possibly contributed to the tragedy. No single factor was, by itself, the sole cause of the incident. Collectively, the factors discussed in this chapter were interdependent and somewhat related, and each influenced the action which took place in a different way. . . .

Plans and Orders

There is substantial evidence that the events at Son My resulted primarily from the nature of the orders issued on 15 March to the soldiers of Task Force (TF) Barker. Previous chapters of this report have described the content of the different orders issued by LTC Barker, CPT Medina, CPT Michles, and the various platoon leaders and have indicated the crucial errors and omissions in those orders. The evidence is clear that as those orders were issued down through the chain of command to the men of C Company, and perhaps to B Company, they were embellished and, either intentionally or unintentionally, were misdirected toward end results presumably not foreseen during the formative stage of the orders.

Peers, *Report,* vol. 1, pp. 8-1–8-2.

The orders derived from a plan conceived by LTC Barker and approved by several of his immediate superiors. There is no evidence that the plan included explicit or implicit provisions for the deliberate killing of noncombatants. It is evident that the plan was based on faulty assumptions concerning the strength and disposition of the enemy and the absence of noncombatants from the operational area. There is also evidence to indicate widespread confusion among the officers and men of TF Barker as to the purpose and limitations of the "search and destroy" nature of the operation, although the purpose and orientation of such operations were clearly spelled out by MACV directives in effect at that time. The faulty assumptions and poorly defined objectives of the operation were not explored nor questioned during such reviews of the plan as were made by MG Koster, BG Lipscomb, and COL Henderson. LTC Barker's decision and order to fire the artillery preparation on portions of My Lai (4) without prior warning to the inhabitants is questionable, but was technically permissible by the directives in effect at that time. The implementing features of that decision were inadequate in terms of reasonable steps that could have been taken to minimize or avoid consequent Vietnamese casualties from the artillery preparation. The orders issued by LTC Barker to burn houses, kill livestock, destroy foodstuffs (and possibly to close the wells) in the Son My area were clearly illegal. They were repeated in subsequent briefings by CPT Medina and possibly CPT Michles and in that context were also illegal.

While the evidence indicates that neither LTC Barker nor his subordinates specifically ordered the killing of noncombatants, they did fail, either intentionally or unintentionally, to make any clear distinctions between combatants and noncombatants in their orders and instructions. Coupled with other factors described in this report, the orders that were issued through the TF Barker chain of command conveyed an understanding to a significant number of soldiers in C Company that only the enemy remained in the operational area and that the enemy was to be destroyed.

13

EUGENE KOTOUC

Testimony to Peers Commission
1970

On the afternoon of March 15, 1968, Lieutenant Colonel Frank Barker convened a meeting to brief his company commanders about the next day's operations. Barker had recently been briefed himself by Colonel Oran K. Henderson, commander of the 11th Infantry Brigade. Captain Eugene Kotouc, a combat intelligence officer and staff aide to Barker, was familiar with Henderson's instructions about the operation. Kotouc also attended Barker's briefing, as did Captain Ernest Medina of Charlie Company. During the Peers investigation, Kotouc was asked about the nature of that meeting and whether Barker had given specific orders about killing civilians.

Q: Were any instructions given concerning the destruction of the village?

A: Yes, sir, there were. Colonel Barker said he wanted the area cleaned out, he wanted it neutralized, and he wanted the buildings knocked down. He wanted the hootches burned, and he wanted the tunnels filled in, and then he wanted the livestock and chickens run off, killed, or destroyed. Colonel Barker did not say anything about killing any civilians, sir, nor did I. He wanted to neutralize the area.

Q: When did he give these instructions?

A: He told me this was what we wanted to do here. When it was, it was prior to the operation. As I recall, it was the night before, or early afternoon before, on 15 March. . . .

Q: Were you familiar with the brigade orders and the division orders concerning the burning and/or destruction of a village or hamlet?

A: I am familiar with Colonel Henderson saying that when we get through down there, there would not be any 48th left, or any place for them to live.

Q: Specifically what did Colonel Henderson say?

A: To the best of my recall, he said that when we get through with that 48th Battalion, they won't be giving us any more trouble. We're going to do them in once and for all. I thought, personally, that was a real fine thing to say.

Peers, *Report,* vol. 2, bk. 16, pp. 11–13, 15, 19.

Q: Did he give any instructions to burn the village?

A: Not to my knowledge, sir.

Q: To destroy the area?

A: Not to my knowledge. . . .

Q: I would like you to again state, to the best of your ability, what Colonel Barker said that you would do with respect to that village, My Lai (4)?

A: Colonel Barker said he wanted the defensive positions destroyed, the bunkers, and the trench work, and the tunnels, if we could find them. He wanted the hootches knocked down, and in a case where they could burn them to burn them up. He wanted the livestock and chickens to be run off, or else destroy them. Run them off and get them out of the area.

Q: Did he say anything about the wells, the water?

A: No, sir, he definitely did not. Colonel Barker never did say anything about polluting wells, or to do anything with the wells, not to my knowledge. I never heard him say anything about that.

Q: Did he say anything about destruction of other villages in later parts of the operation?

A: He spoke in general terms, and frankly, the village My Lai (4) was not given a whole lot—we did not talk constantly about My Lai (4). The operation was not for My Lai (4) per se. It was for the area there. The reason we went to My Lai (4) was because that is where we thought the headquarters and the two battalions were. They were to sweep through and move, and as my memory serves me, they moved through the north-northeast. There would be a blocking company up there. The idea was that anything that could be used in the defensive position or blocking position to give aid and cover was to be destroyed. . . .

Q: Were the company commanders given to understand that they had a chance to encounter the 48th VC Battalion in and around My Lai? Were they given to understand that they had a chance to trap the 48th VC Battalion in My Lai and around there?

A: That was the hope of the whole thing, that we could get them into there, pinch them in, and do battle with them right there, and that Michles' company would be in a position where they could not get away if they ran. They would just take off, they would scatter to the winds. It was the whole picture that we could suck them in, pinch them in, and destroy them.

Q: In the discussion of the plans, you planned to destroy the base of operations in and around My Lai (4). Could that have been given in such terms that the company commanders got the idea that this should include that part of the civilian population which supported the VC?

A: To do away with them, sir?

Q: Yes.

A: No, I do not believe so, sir. It certainly wasn't—

Q: (Interposing) I understand you testified this certainly was not said specifically.

A: Sir, I do not think it was inferred, sir.

Q: Included in the field order, were there instructions, specific instructions, to watch out for them the best you could, and to see that no harm came to noncombatant civilians. Was that discussed on the 15th?

A: Sir, I do not think it was referred to, sir. . . .

Q: We will come back to the civilians. But going back to the planning stages, the directive stages of the operation. Were there not any considerations given to the care and handling of the civilians? If the order of Colonel Barker had just been issued to destroy the place, it would have been obvious that you were going to have—that there was going to be a refugee problem of considerable magnitude. Was any consideration given to that?

A: Yes, sir, there was. It was an SOP for the unit when civilians were present in the area, if a fight was going on, the civilians would be taken and moved down the road, in this case to Quang Ngai City. We told them to please get out of the area, and go down to Quang Ngai City. We had been telling those people for a long time to get out of there and live in Quang Ngai City where they had refugees places. It was a policy, and I personally saw it happen when they take the people and just move them through the lines and go down the road. Normally, with all the shooting, they were scared, and they would go on down the road. We did not appoint, and say: "You take civilians down the road." We did not do that.

Q: Were there any instructions issued by Colonel Barker or Colonel Calhoun, or anybody, to take [. . .] adequate provisions for marching these civilians out of the area, moving them down in an orderly fashion to Quang Ngai City, and assuring that they were adequately cared for?

A: There was not, because I brought it up myself. I wanted to know what kind of release point they wanted for the civilian population. And they said to do it like normally, and move them down the road.

ERNEST L. MEDINA

Testimony to U.S. Army CID
1969

Several hours after Lieutenant Colonel Barker's briefing, on the evening of March 15, 1968, Captain Ernest Medina held a meeting with the troops of Charlie Company, at which he explained the nature and objectives of the scheduled assault on My Lai. Exactly what he told his troops that night remains a matter of considerable controversy. Medina later denied giving orders to kill every living thing—men, women, children, and animals—at My Lai. Most of the troops at the meeting remember instructions to that effect, but a few members of the company are confident that Medina did not issue such orders. In any event, many of the troops came away from the briefing convinced that whether or not they encountered North Vietnamese or Vietcong troops, the operation was going to be a bloody affair, a chance to avenge the deaths of their comrades. During the Peers investigation into the My Lai massacre Medina was asked if he had told his platoon leaders to kill all of the My Lai villagers.

Q: Did you brief your company?
A: Yes, sir, I did.
Q: In total?
A: Yes, sir, I did.
Q: What did you tell them?
A: Well, I told them we were going into the Pinkville and I gave them a brief rundown on the background information, what I knew, and, of course, they were aware of the other companies that had gone in there and had taken a number of casualties, and I told them that the LZ preparation would be put onto the village where the LZ would be. I kind of sketched it out on the ground for them and gave them a general rundown of how we would go in there, the elements that would be pushing through first and move the people, and the search elements would go in afterwards and make their search, and then we would marry up and both companies would ring up for the night or our company would ring up for the night in a defensive position to the west. . . .

CID Deposition Files, My Lai Investigation, CID Statement, file no. 70-CID011-00013, U.S. Army Crimes Records Center, Fort Belvoir, Virginia, pp. 259–60.

Q: Did you order the burning of the village?

A: Yes, sir, I told my company that we were authorized to burn the village.

Q: Who did you assign to perform this mission?

A: That was up to the search element, sir. Once they finished searching they would burn the hootches and whatnot, sir.

Q: Did you order the destruction of the inhabitants?

A: No, sir, I did not. That is, if there was any, sir, I did not.

15

MAX D. HUTSON

Testimony to U.S. Army CID

1969

Max D. Hutson was the weapons squad leader with the 2nd Platoon of Charlie Company.

The night prior to the mission ... CPT Medina called the company together and explained the mission to us. He stated that My Lai #4 was a suspected VC strong hold and that he had orders to kill everybody that was in the village. We did not expect to find anyone in the village, and when we did, we did as ordered.

CID Deposition Files, My Lai Investigation, CID Statement, U.S. Army Crimes Records Center, Fort Belvoir, Virginia, p. 261.

GREGORY T. OLSEN

Testimony to U.S. Army CID

1969

Gregory Olsen, a devoutly religious Mormon from Portland, Oregon, was with Lieutenant William Calley's 1st Platoon during the assault on My Lai.

Q: Prior to the assault on My Lai (4) did the Company receive a briefing?

A: Yes the company did. The briefing was given by CPT Medina and I attended the briefing. At the time everybody was down in the dumps, because just previous due to various operations in the past few weeks we had lost about 25 men. 7 of them had been killed and the rest wounded. The briefing was given at LZ Dottie, where CPT Medina, drew a map on the ground and explained the entire procedures. We had instructions to shoot on sight any military age male, running from us, or shooting at us. We were then told, that we are to clear all the people out of the village. He (CPT Medina) did not say anything about the disposition of the people that we had or would clear out of the village. We were told, to destroy all the food supplies and the animals in the area. I do not remember if in the initial briefing we were told to burn all the huts. CPT Medina made the statement that we owed the enemy something. The troops had a feeling that they should revenge their fallen comrades.

Q: Did CPT Medina ever order during the aforementioned briefing to kill all the inhabitants of the village? With all the inhabitants I mean also women and children.

A: Negative. He did not. CPT Medina, was in my opinion an outstanding Commander. He was always concerned with the welfare of his men. Sometimes we did things the hard way, but in the end it was always the best for us. CPT Medina would never have given an order to kill women and children.

Q: Was LT Calley present during the briefing?

A: I assume he was.

Q: Did you attend a briefing on the operation My Lai (4) by LT Calley?

CID Deposition Files, My Lai Investigation, CID Statement, U.S. Army Crimes Records Center, Fort Belvoir, Virginia, p. 364.

A: I only remember that he told us on which helicopters we were supposed to go on. I do not remember LT Calley giving us a specific briefing on My Lai (4) after CPT Medina had briefed us. I do not remember who my squad leader was during the Pinkville Operation. It is quite a long time ago.

17

HARRY STANLEY

Testimony to U.S. Army CID
1969

Harry Stanley was a native of Gulfport, Mississippi. On the day of the My Lai operation, he was assigned to the 1st Squad in Lieutenant William Calley's 1st Platoon.

The day before the attack, I attended a company briefing which was conducted by Captain Medina. This briefing was attended by all of the Platoon Leaders and Platoon Sergeants and by most of the men in the company. Captain Medina told us that the intelligence had established that MyLai (4) was completely enemy controlled. He described the formations we were to use the following day and told us to carry extra ammunition. He ordered us to "kill everything in the village." The men in my squad talked about this among ourselves that night because the order to "kill everything in the village" was so unusual. We all agreed that Captain Medina meant for us to kill every man, woman, and child in the village. This was the only briefing. The following morning the Platoon Leaders and Platoon Sergeants got us in order for the helicopters, but did not brief us as such.

CID Deposition Files, My Lai Investigation, CID Statement, U.S. Army Crimes Records Center, Fort Belvoir, Virginia, pp. 431–32.

RONALD L. HAEBERLE

Testimony to U.S. Army CID

1969

Ronald Haeberle, a native of Cleveland, Ohio, was a combat photographer. On the morning of March 16, 1968, he was assigned to Charlie Company and accompanied the troops to My Lai.

Q: Have you heard that Task Force Barker was given the mission to destroy the village in question and to kill all the inhabitants?

A: I heard something to that effect. It was general talk amongst the soldiers. I do not [know] if this order in fact was given. The soldiers were saying that all the inhabitants were communists or sympathizers or Viet Cong. I did not attend any briefings for the companies involved. I joined Company C on the morning of the assault. During the helicopter ride to the village nothing was mentioned of killing all the inhabitants. The talk about killing the inhabitants started on the ground. I remember vaguely that either Cpt Medina the CO of Company or the MI man ("Bull") explained to the Vietnamese interpreter (Sgt Phu) why the killing of the inhabitants had to be done. I had also heard the day before the assault, that the villagers were instructed to leave the village. How they were told to leave I do not know. I do not know if there was an artillery barrage on the village prior to our landing there.

CID Deposition Files, My Lai Investigation, CID Statement, U.S. Army Crimes Records Center, Fort Belvoir, Virginia, p. 88.

HERBERT L. CARTER

Testimony to U.S. Army CID

1969

Herbert Carter, who served as a "tunnel rat" with Charlie Company, became the only U.S. casualty at My Lai when he accidentally shot himself in the foot.

During March 1968, we were moved to LZ Dottie. We stayed there a day or so and then we were told that we were going to an area we know as Pinkville. The night before the operation, Captain Medina gave the unit a pep-talk and a briefing. The briefing was the usual: equipment to take, what order we would go in, etc. The pep-talk was unusual. He said "Well, boys—this is your chance to get revenge on these people. When we go into MyLai (4), it's open season. When we leave, nothing will be living. Everything is going to go." He also said to level the village.

After this briefing, Calley told me to double my ammunition supply. This wasn't at a briefing: he came around to all the bunkers.

When I left Medina's briefing I knew it was going to be a slaughter of civilians in the village the next day. Stanley and I talked about this before MyLai (4) and we agreed then that it would be a slaughter.

Now I know I am going to say things against friends of mine, but this is the time for the truth. Take Mitchell, I've been out with him socially a couple of times, but in MyLai (4) he killed people he had no reason to kill. He murdered them. Meadlo was my friend too, but he murdered people too. I got along with Calley until MyLai (4), but after he killed all those people I couldn't take any more of that.

CID Deposition Files, My Lai Investigation, CID Statement, U.S. Army Crimes Records Center, Fort Belvoir, Virginia, p. 145.

ROBERT W. T'SOUVAS

Testimony to U.S. Army CID

1969

Robert T'Souvas was a member of the 3rd Platoon of Charlie Company on March 16, 1968.

Q: On the night before the assault into My Lai (4) CPT Medina gave a pre-assault briefing to the company. Relay to me your own words that CPT Medina told the troops.

A: I do not know if it was right or wrong, but we were briefed about Pinkville and were told that it was heavily populated with Vietcong and North Vietnamese Army. We were also told that all the people in the Hamlet were VC sympathizers. Our mission was supposed to be a search and destroy mission and we were told so.

Q: Did CPT Medina at any time, tell the members of C Company 1/20th Infantry that they should kill all the animals, kill all the inhabitants, and shoot on anything that moves?

A: To my knowledge he did.

Q: Did the Company take this as having to shoot all the women and children and burn all the hootches in the Hamlet?

A: Our orders were to kill everything in the village and to burn everything.

Q: Did CPT Medina at anytime during the briefing tell the Company that when he comes through the "Ville" the next day, that all he wants to see living are members of Company C?

A: No, I did not hear that.

Q: How did the Company react to the briefing of CPT Medina?

A: I cannot express my feelings for the rest of the company, except my own. I don't know how they felt, myself I felt strange in a way but we were told Pinkville My Lai (4) (5) (6) were heavily populated with Viet Cong and VC sympathizers. We were also told that most of the snipers that had attacked our company had came [sic] from that area.

CID Deposition Files, My Lai Investigation, CID Statement, U.S. Army Crimes Records Center, Fort Belvoir, Virginia, p. 7.

21

THOMAS R. PARTSCH

Journal Entry

March 15, 1968

Immediately after the briefing, Thomas R. Partsch of Charlie Company made the following entry in his personal journal.

<u>Mar 15 Fri.</u> got up at 5 a.m. ate chow had police call after that. Had garbage detail sure get a kick out of that. Brought mess garbage down and they even ate that. Just sitting around now cleaning weapons and took another bath at the stream. Had meeting [w]hole company said we are going to really hit something tomorrow going to hit 4 places its a hot place.

Peers, *Report,* vol. 4, exhibit M-85, p. 299.

22

ROBERT E. MAPLES

Testimony to U.S. Army CID

1969

Robert Maples was a machine gunner with Charlie Company on the morning of March 16, 1968.

I do remember taking part in the operation. During the operation CPT Medina, was the commanding officer, LT Calley was the platoon leader, SGT Cowan was the platoon SGT and SGT Mitchell was the squad leader of 1st Platoon, 1st Squad. The night before the mission the company was briefed by Medina. During the briefing Medina told the company there

---

CID Deposition Files, My Lai Investigation, CID Statement, file no. 69-CID011-00073, U.S. Army Crimes Records Center, Fort Belvoir, Virginia, p. 13.

was supposed to be nothing in the village but the enemy, that we were to sweep through the village and kill everything that was in there. He also told us to put the dead animals in the wells. He did not leave the impression with me that he meant to kill the women and children however this is what did happen.

23

NGUYEN DINH PHU

Testimony to Peers Commission
1970

Nguyen Dinh Phu, a member of the South Vietnamese army, was assigned as an interpreter to Captain Ernest Medina. Phu did not attend the meeting in which Medina briefed Charlie Company, but he did visit with several squads after the meeting on the eve of the assault on My Lai.

A: ... I did not participate [in the pre-operation briefing]. I was not present at any meeting on the 15th between the soldiers, or the officers and the commanders, that Captain Medina held with his subordinates. Normally, Captain Medina conducted a meeting at night, frequently late at night, with soldiers before they conducted an operation. I last saw Captain Medina about 8 o'clock in the evening. After that, I went out into the village for recreation. That is different from what I said I recalled a few minutes ago.

Q: Well, I would like to come back to the question, though, as to whether or not you heard Captain Medina or whether you had heard anything about any instructions that had been issued, whether you heard it or whether it was hearsay, as to what they were to do the following day?

A: Sir, the time of my departure to go to the village may have been earlier, it may have been around 6 o'clock. This is material in that after I returned it was probably around 8 o'clock. It could have been later, but I did go to the village and return. During the time I was gone, there

Peers, *Report*, vol. 2, bk. 32, pp. 5–6.

was a meeting held. Upon my return, I met a large group of soldiers. Many of the soldiers were drunk, a large number. One of the soldiers told me that tomorrow they would go on an operation and they would kill women, children, cattle and everything. I do not recall the name of that soldier. Only one soldier told me.

Q: I'm interested in the soldiers being drunk. What were they drinking? Were they drinking beer or what were they drinking, and what was their condition?

A: They were drinking beer and whiskey too, on the bunker line. They had bottles of whiskey and cans of beer in the bunkers on site.

Q: Was this just one squad, or were all of the men drinking?

A: The bunkers were very near. I visited two or three squads. Virtually all of them were drinking as if they were having a party; a lot more than normal, a lot more than usual.

Q: What did you think about what this soldier is telling you, that tomorrow they were going to kill VC, and women and children as well?

A: I was very surprised. "Are you joking with me?" I asked the man. That person says, "I'm not joking, that's the truth." That person was drunk and I discredited it to some extent because he was. After that, I drank with them.

Q: How long did this drinking go on before the men went to bed?

A: About midnight. It was very late. I also drank until I was drunk, but I would estimate around midnight.

Q: Was Captain Medina or any of the officers around during this time?

A: No, sir.

Q: Did the men who were on the bunker line that night later on move away from the bunkers, and somebody else take over the bunker security?

A: No, sir, that's where they lived. They stayed there.

Q: Do you remember anything more about that night that may have a bearing upon our investigation?

A: This was the last day of standdown. Normally, they did do a good deal of drinking during the days of standdown. But, normally the night before they went on an operation, they didn't drink excessively. So this was, in my experience, somewhat unusual. I do not recall any other incident about that night that might be significant.

24

MICHAEL TERRY

Testimony to U.S. Army CID
1969

Michael Terry was a fire team leader in the 3rd Platoon of Charlie Company on March 16, 1968.

Q: Could you describe what had happened during previous operations in the area?

A: . . . A couple of days before this Pinkville incident we had had one guy killed and about three or four were wounded with legs blown off or something similar to that. Before this time we had had the same thing and the men hadn't had much action or anything, no way to fight back, which made the men unhappy probably. The last time was the last straw and the men, they had a meeting together and talked to the captain and some of them broke down and cried and things like that, and they asked when they would be able to fight, to let go with their feelings or things like that; and in this meeting the captain said that, words to the effect that we were going down into Pinkville. We had been there before and had lost a couple of guys and knew that it was a hot war and we had a mission down in Pinkville.

He said he knew we would be able to fight then. Now I noticed, I paid attention particularly to see if he was going to come out and say these words. A couple of times they asked if they could shoot anything they saw.

Q: Who did they ask?

A: The captain.

Q: Captain who?

A: CPT Medina. They asked him and he never said that they could, but the idea, I could tell the idea that these men got was that they could, that nobody in there was friendly, nobody in the village. Now then, another thing to preview this. These men — I mean a lot of GI's, and these in particular in this case, they don't feel empathy with the Vietnamese when they are thrown together, so they don't feel any empathy toward them and they treat them as if they weren't human beings at times.

CID Deposition Files, My Lai Investigation, CID Statement, U.S. Army Crimes Records Center, Fort Belvoir, Virginia, p. 51.

DENNIS CONTI

Testimony to Peers Commission
1970

After Captain Medina's briefing, Lieutenant William Calley convened a similar briefing for his platoon. Dennis Conti was a member of Calley's platoon on March 15, 1968.

Q: What kind of a company commander was Captain Medina?

A: A good company commander.

Q: Did he take care of his men?

A: Yes, he watched out for his men. I believe he did the best for his men and the job we had to do.

Q: Did anybody else join in the briefing of the company that night?

A: We had one by Lieutenant Calley.

Q: What did Lieutenant Calley say to you?

A: He told us relatively the same thing. He said that when we go in, any men there, or something to the effect that any men found there will have a weapon, any women will have a pack, any cattle is VC food, and to destroy it.

Q: What did you understand from this with regard to the Vietnamese residents of the village?

A: From what he said, there were no if's, and's or but's. They will have a weapon. Whether they did or not, his tone of voice, the way he said it, that's the way I interpreted it. Usually they'll say, from my experience: "If they have a weapon," not, "They will," in a strong tone of voice.

Q: What do you think he meant by this?

A: I figured he was going in and, I don't know, at the time I figured myself that there was going to be strong resistance. I figured there would be men in the village, but I figured the men would be armed, and I figured that they would be supported by the women. That's the only thing I interpreted from it. Like I said, I interpreted that there would be strong resistance. . . .

Q: What was the mood of the men after the briefing?

Peers, *Report,* vol. 2, bk. 24, pp. 27–29.

A: I think we were "psyched up," ready for battle more or less. But, like I say, we were ready to meet a foe of equal military strength, if not greater. And we prepared to give our best.

Q: How about Lieutenant Calley?

A: He seemed the same way but a little more, from the speech he gave, a little more "psyched up."

Q: Did he say anything about the children?

A: He said something to the effect that they would be future VC, or something like that. He said something about the children, something like that, but I can't remember his exact words.

Q: What do you think he meant by that?

A: I assume he meant the children were the same as the mothers and fathers, they were VC. That's it, no two ways about it.

Q: Do you think he meant by that that they would be shot, men, women, and children?

A: I don't know. I think, at the time, that's the way I interpreted it.

Q: Did the other men in the platoon get the same impression?

A: I don't know. I couldn't speak for anybody else.

Q: Do you recall a memorial service being held before the briefing?

A: Yes.

Q: Could you tell us about that?

A: They had a memorial service for three of the men who were killed in the minefield. One was Rocker, I think there was a Bell, and the other kid was in my squad and I can't think of his name.

Q: Wilson?

A: Bobby Wilson.

Q: When was the memorial service held with reference to the briefing?

A: I think just prior to it.

Q: The same afternoon?

A: Yes.

Q: When you say every man has a weapon, every woman has a pack, and the children will grow up to be VC, your impression was, and I want to make sure I'm correct in this, you're going to go in there and you're going to kill everybody. Is there any other possible interpretation?

A: Like I said, at the time we believed we were going to hit a foe. We figured we were going to hit the village, and the village was going to be heavily armed. The way I interpreted it was, if they were in the way, kill them. You know, like, if you're pinned down, and there's four or five heavily armed men, and they have a woman in front of them, that's the way I interpreted it. But I didn't believe he meant just to go in and massacre them. That's the way I interpreted it.

5

The Assault on My Lai

In the following selections, American and South Vietnamese participants and witnesses describe what happened on the morning of March 16, 1968, in My Lai village. Their testimony raises a number of compelling moral and psychological issues. As you read the documents, you will discover that some of the troops, without question, obeyed Lieutenant William Calley's order to kill civilians. There is little doubt about the reality of such orders, but in your opinion, are the troops who obeyed guilty of murder and war crimes? Other members of Calley's platoon knew immediately that such orders were illegal, and they refused to obey. But even those who disobeyed Calley's orders did not actively engage in stopping the killing. They simply refused to participate. To what extent might they be accomplices to the slaughter? Also, they did not report the alleged war crimes to higher authorities as required by the Geneva conventions. Did such behavior render them guilty of participating in the cover-up of the massacre? Only Hugh Thompson, a helicopter pilot, actively worked to protect Vietnamese civilians from the American troops. He even threatened to fire on his own men if they continued the slaughter. Was he right or wrong in his actions? In addition, Thompson was the only American at My Lai that day to report the massacre to superior officers.

The question of "mercy killings" at My Lai also raises difficult moral issues. Some troops at My Lai refused to participate in the initial slaughter of civilians, but after the massacre, as they patrolled the village, they came across badly wounded people whose chances of survival were nonexistent. Medical care was not being given to wounded civilians, and these soldiers were certain that the Vietnamese would not survive and were therefore suffering needlessly. So the troops "finished them off." In implementing such so-called mercy killings, were these troops also guilty of war crimes and murder?

Finally, the My Lai testimony raises the issue of rape as it occurred at My Lai and throughout Vietnam. Ever since the My Lai massacre was first exposed to public scrutiny, some people have defended Calley on the grounds that Vietnam was a "different" kind of war, that because American soldiers often could not distinguish between civilians and enemy troops, they had to attack noncombatant Vietnamese to protect themselves. According to this argument, the troops who massacred civilians were not guilty of war crimes. But that argument does not take into account the rape of dozens of Vietnamese women and young girls at My Lai. In what way can rape be considered self-defense?

26

DENNIS CONTI

Testimony to Peers Commission
1970

Dennis Conti, a grenadier and mine sweeper with Lieutenant William Calley's 1st Platoon, gave the following account of the massacre to Peers investigators.

Q: You got to the landing zone [west of My Lai 4] early in the morning?
A: We hit the landing zone and I had the mine sweeper. We jumped off—
Q: (Interposing) You were with Meadlo?
A: Yes, I was in the first chopper. And I think I found Lieutenant Calley, and joined up with him. And at the time there was, to the right, along the tree line, a villager with cattle moving out. And I heard somebody yelling: "They're running away; they're running away." And they opened up with a M-60. So I was in the open, and it was in my line of fire, and I fired at the cattle. I missed the cattle, they were too far out of my range. Then, after that, we got up, and we moved into the village, caught a trail and went into the village. I saw Sergeant Bacon and his squad, and I stopped and talked to them for a while.
Q: Sergeant Bacon was on the left, was he?

A: I couldn't tell you. All I know is I met up with him, and then he chewed me out for not being with the CP, and told me to get back with them. I moved up through the village. On the way, I guess, there was a few people killed there, there were bodies there. I moved up, and I met Lieutenant Calley again in the CP group. When I got there, we were told to round the people up. So myself and Meadlo, I had the mine sweeper and I couldn't do anything, so most of the guys were rounding them up, and bringing them to me and Meadlo. We herded them all together, pushed them out. He said: "Bring them out into the rice paddy."

Q: Mr. Conti, I show you the aerial photograph that you annotated this morning, which was admitted into evidence as Exhibit P-133. Could you indicate to us on the photograph the part of the village where you were gathering the people together?

A: Around in here.

Q: Just north of point 4?

A: Yes, right around there. They were bringing people out, and then we pushed them out into the rice paddy, onto the dike there. And, like I said, we pushed them out there. Meadlo and myself, we watched them. While we were watching, a little kid came running out up here, and I went up to investigate. I told him to watch the people. There were a woman and a baby about 4 years old, who were walking, and an older woman, who I assumed to be a grandmother or something. I rounded them up, brought them back down to Meadlo, and we stood around them for a couple of minutes talking. Lieutenant Calley came back, and said: "Take care of them." So we said: "Okay." And we sat there and watched them like we usually do. And he came back again, and he said: "I thought I told you to take care of them." I said: "We're taking care of them." And he said: "I mean kill them." So I looked at Meadlo, and he looked at me, and I didn't want to do it. And he didn't want to do it. So we just kept looking at the people, and Calley calls over and says: "Come here, come here." People were right around here . . . and we were on the other side. Then he said: "Come on, we'll line them up here, we'll kill them." So I told him: "I'll watch the tree line. There's a tree line over here." I had the M-79, I figured that was a good excuse. I had the M-79. I didn't want to waste my ammo. I'll watch the tree line. So I watched the tree line. Myself and Meadlo were over here, and I think he told them [another group of soldiers]: "When I say fire, fire at them." Then they opened up, and started firing. Meadlo fired a while. I don't know how much he fired, a clip, I think. It might have been more. He started to cry, and he gave me his weapon. I took it, and he told me to kill them. And I said I wasn't going to kill them. At the time, when we were talking, the only thing left was children. I told Meadlo, I said: "I'm not going to kill them. He [Calley] looks like he's enjoying it. I'm going to let him do it." So, like I said, the only thing left was children. He [Calley] started killing the children. I swore at him. It didn't do any good. And that was it. They were all dead. He turned around, and said: "Okay, let's go." We turned around, and walked away. Somebody yelled: "They're getting away." I turned around and looked, and there was a group of people running over here toward the tree line which was my responsibility. Lieutenant Calley yelled: "Get them; kill them." So I gave them a chance to get in the tree line, and I opened up with four rounds. We didn't bother to investigate. We left. When we started to leave they took off ahead of me, because I didn't want to go with them.

Q: Did this occur in the vicinity of point 4?

A: Yes, this is all right here on the trail. Like I said, I didn't want to walk with them. I was walking around in a circle myself. They went away, they went up the trail. I walked up the trail, and then I wandered back to the village. I don't know exactly where I went. And I ended up here at point 5, by this house here. I stopped, I talked to a couple of guys who were there. I don't remember who they were. At the time I saw somebody over here firing into the ditch, this ditch over here. So I thought maybe we have been hit. I didn't know, so I went over to investigate, to see if anybody needed help. I walked over there. I walked over to the ditch. As I walked up to it, Lieutenant Calley and Sergeant Mitchell were firing into the ditch. I looked into the ditch, and I saw women, children, and a couple of old men, just regular civilians. I saw a woman get up, and Calley shot her in the head. She went back down. I didn't feel like watching any more, so I turned around and walked away.

Q: About how many do you think were in that ditch, this last one, at point 6?

A: They were all along the ditch. It could have been 40, maybe more, maybe less. I don't know. I figure around 40.

Q: How about the first group you saw killed near point 4.

A: About 40.

27

HERBERT L. CARTER

Testimony to U.S. Army CID

1969

Herbert Carter, a "tunnel rat" who served in the 1st Squad of Lieutenant William Calley's 1st Platoon, refused to fire at civilians at My Lai. During the operation, his pistol accidentally discharged, wounding him in the foot. Following is Carter's account of what happened on the morning of March 16, 1968.

We were picked up by helicopters at LZ Dottie early in the morning and we were flown to My Lai (4). We landed outside the village in a dry rice paddy. There was no resistance from the village. There was no armed enemy in the village. We formed a line outside the village.

The first killing was an old man in a field outside the village who said some kind of greeting in Vietnamese and waved his arms at us. Someone—either Medina or Calley—said to kill him and a big heavy-set white fellow killed the man. I do not know the name of the man who shot this Vietnamese. This was the first murder.

Just after the man killed the Vietnamese, a woman came out of the village and someone knocked her down and Medina shot her with his M16 rifle. I was 50 or 60 feet from him and saw this. There was no reason to shoot this girl. Mitchell, Conti, Meadlo, Stanley, and the rest of the squad and the command group must have seen this. It was a pure out and out murder.

Then our squad started into the village. We were making sure no one escaped from the village. Seventy-five or a hundred yards inside the village we came to where the soldiers had collected 15 or more Vietnamese men, women, and children in a group. Medina said, "Kill everybody, leave no one standing. Wood was there with an M-60 machine gun and, at Medina's orders, he fired into the people. Sgt Mitchell was there at this time and fired into the people with his M16 rifle, also. Widmer was there and fired into the group, and after they were down on the ground, Widmer passed among them and finished them off with his M16 rifle. Medina, himself, did not fire into this group.

CID Deposition Files, My Lai Investigation, CID Statement, file no. 69-CID0011-00074, U.S. Army Crimes Records Center, Fort Belvoir, Virginia.

Just after this shooting, Medina stopped a 17 or 18 year old man with a water buffalo. Medina said for the boy to make a run for it—he tried to get him to run—but the boy wouldn't run, so Medina shot him with his M16 rifle and killed him. The command group was there. I was 75 or 80 feet away at the time and saw it plainly. There were some demolition men there, too, and they would be able to testify about this. I don't know any other witnesses to this murder. Medina killed the buffalo, too.

Q: I want to warn you that these are very serious charges you are making. I want you to be very sure that you tell only the truth and that everything you say is the truth?

A: What I have said is the truth and I will face Medina in court and swear to it. This is the truth: this is what happened.

Q: What happened then?

A: We went on through the village. Meadlo shot a Vietnamese and asked me to help him throw the man in the well. I refused and Meadlo had Carney help him throw the man in the well. I saw this murder with my own eyes and know that there was no reason to shoot the man. I also know from the wounds that the man was dead.

Also in the village the soldiers had rounded up a group of people. Meadlo was guarding them. There were some other soldiers with Meadlo. Calley came up and said that he wanted them all killed. I was right there within a few feet when he said this. There were about 25 people in this group. Calley said when I walk away, I want them all killed. Meadlo and Widmer fired into this group with his M16 on automatic fire. Cowan was there and fired into the people too, but I don't think he wanted to do it. There were others firing into this group, but I don't remember who. Calley had two Vietnamese with him at this time and he killed them, too, by shooting them with his M16 rifle on automatic fire. I didn't want to get involved and I walked away. There was no reason for this killing. These were mainly women and children and a few old men. They weren't trying to escape or attack or anything. It was murder.

A woman came out of a hut with a baby in her arms and she was crying. She was crying because her little boy had been in front of her hut and between the well and the hut and someone had killed the child by shooting it. She came out of the hut with her baby and Widmer shot her with an M16 and she fell. When she fell, she dropped the baby and then Widmer opened up on the baby with his M16 and killed the baby, too.

I also saw another woman come out of a hut and Calley grabbed her by the hair and shot her with a caliber .45 pistol. He held her by the hair for a minute and then let go and she fell to the ground. Some enlisted man standing there said, "Well, she'll be in the big rice paddy in the sky."

Q: Do you know any witnesses to these incidents?

A: Stanley might have [seen] the one Calley killed. There were a lot of people around when Widmer shot the woman with the baby. I can't definitely state any one person was there, but there were a lot of people around.

I also saw a Vietnamese boy about 8 years old who had been wounded, I think in the leg. One of the photographers attached to the company patted the kid on the head and then Mitchell shot the kid right in front of the photographer and me. I am sure the boy died from the fire of Mitchell.

About that time I sat down by a stack of dying people and Widmer asked me if he could borrow my caliber .45 pistol and finish off the people. I gave him my pistol and he walked in among the people and would stand there and when one would move, he would shoot that person in the head with the pistol. He used three magazines of caliber .45 ammunition on these people. These were men, children, women, and babies. They had been shot by machinegunners and riflemen from Company C, 1/20th Infantry. This was at a T-junction of two trails on the outskirts of the village. I got my pistol back from Widmer and holstered it again.

Q: How many people do you figure Widmer finished off when he used your pistol?

A: I know he shot some twice, so I figure he shot fifteen or so with my pistol. I know he shot one guy in the head and I imagine that was where he was shooting them all.

Q: What happened then?

A: We went on through the village and there was killing and more killing. I was with Stanley, mainly. I sat down with Stanley and Widmer came up again and asked to borrow my pistol again. I gave it to him. I saw a little boy there—wounded, I believe in the arm—and Widmer walked up close to the kid and shot him with my pistol. Widmer said something like, "Did you see me shoot that son of a bitch," and Stanley said something about how it was wrong. My gun had jammed when Widmer shot the kid. As far as I could tell, the kid died as a result of this gunshot. Then Widmer gave me my pistol back and walked off. I was trying to clean it when it accident[al]ly went off and I was shot in the left foot. Stanley gave me medical aid and then the medics came. Medina and some of the command group came up and then I was flown out in a helicopter. The next day the medics brought Meadlo into the hospital. He had stepped on a booby-trap and had lost his foot. He said he thought God might be punishing him for what he had done in My Lai (4)....

Q: Did you murder anyone in Vietnam?

A: The only people I killed in Vietnam I killed in combat. I didn't kill any women or kids or unarmed persons at all, ever.

Q: How many people do you think were killed in My Lai (4)?

A: There were more than 100, but I couldn't tell you accurately how many people were killed. I don't believe there were any people left alive.

28

MICHAEL TERRY

Testimony to Peers Commission
1970

Private Michael Terry, of the 3rd Platoon, admitted to what he considered "mercy killings" of wounded civilians at My Lai.

Q: ... Did you fire on wounded civilians?

A: O.K., I will explain. There were some people—you remember the area I told you about where they were rounded up and shot down?

Q: Yes.

A: When we came up to that area, our platoon came up to that area, we were right in that area and there were some people there that we couldn't tell if they were dead or not. Well, a couple of them must not have been dead because they kept whipping around and their heads had been blown off or their brains were sticking out and it was a sickening sight, their limbs were just wiggling, and I remember we shot a couple of those.

Q: Why?

A: Our intent in doing this was—I mean there was no way that they could live and so we just tried to make it faster for them.

Q: How do you know that?

A: When half their head is missing or their brains sticking out, and for one thing, we didn't judge them in any way, but no helicopters were called in there to evacuate the wounded or anything and they just left them laying there and they would just move on ahead.

Q: What was your duty assignment?

A: My duty assignment was team leader.

Q: Who was with you when this incident took place?

A: This specific one I just mentioned?

Q: Yes.

A: Bill Doherty. . . .

Q: Was he a subordinate of yours?

A: Yes, there was about three guys under me.

Q: Did you order him to do this?

A: I don't think you could say I ordered him to do it in any way. I mean we just kind of did it ourselves, you know, on our own.

Q: Did you kill these civilians?

A: Did we kill them? I don't know if you could construe it to say that.

29

ROBERT E. MAPLES

Testimony to U.S. Army CID

1969

Robert Maples, a machine gunner in the 1st Platoon, refused to obey Lieutenant William Calley's direct order to shoot civilians.

Q: This investigation concerns an assault upon a village by the name of My Lai (4), in Quang Ngai Province, Vietnam, that was part of the operation you refer to as "Pinkville." Do you recall participating in this operation and if so will you state what knowledge you have of it?

CID Deposition Files, My Lai Investigation, CID Statement, file no. 69-CID011-00073, U.S. Army Crimes Records Center, Fort Belvoir, Virginia, pp. 13–14.

A: I do remember taking part in the operation.... I do not recall what time we landed or what side of the village of My Lai (4) we landed on. I was in the first lift in the fourth ship and with me was SGT Cowan and my machine gun helpers but I do not recall which ones but it might have been Stanley, Bergthold or Bryant. After we landed the whole company got on a line and started a sweep through the village. Prior to starting through the village I spotted someone off to my right front and I fired on this individual with my machine gun. I was standing while firing and could see that I did not hit him and he got out of my sight somewhere but I do not know where he went. As we went through the village I saw our troops shooting the villagers but I do not know who was actually shooting. I do recall . . . Bergthold shooting a man as the man came out of a hootch. I remember seeing the top of the man's head fly off as Bergthold shot him with .45 pistol. There were other troops ahead of me and as I went through the village I saw dead persons lying in all positions everyplace. I would estimate that I saw about 25 persons dead as I went through the village. As I reached the edge of My Lai (4) I saw LT Calley herding a group of about 15 persons into a hole or a crater. After these persons were into the hole I saw Calley open fire on them with an M-16. I recall that one of the women had been wounded prior to this time and she came up to me and showed the wound. She had been shot in the left arm but there was nothing I could do for her because she was being pushed ahead by two or three other GIs. I did not go and look into the hole but I stayed in the immediate area for some time and none of the persons ever came out of the hole. I believe that a bub[b]le chopper pilot witnessed this incident shortly thereafter he landed and I heard that he told whoever he talked to on the ground that there was still someone alive in the hole. I also heard that the pilot was mad about the incident. I heard that somebody went back to the hole and finished killing those that were not already dead. . . .

Q: Were others than Calley firing into person[s] in the hole?

A: Yes but I do not know who they were. There was either two or three. I do now remember that Medloe [Meadlo] was one of those firing and he was crying at the same time. I know that he or the others did not want to kill those persons. This is not true of Calley because he seemed to want to kill. I do also believe that Calley called Stanley to where he was at the hole to act as interpreter but apparently he did not get any information. After this incident or after lunch we moved on to the next village.

Q: Did you see any animals killed in My Lai?

A: Yes all animals were killed.

Q: Was all the persons in the village also killed?

A: I did not see anyone alive when we left the village.

Q: How many weapons or other military equipment was captured at My Lai?

A: None that I saw.

Q: Were any prisoners taken at My Lai?

A: None that I saw. Later that afternoon some 6 or 8 prisoners were taken at another village. Even later I saw these prisoners being [interrogated] by the Vietnamese National Police. There must have been 2 or 3 police and they walked down the line of prisoners and said "You are VC" and then the other police would take these persons right by the company CP and shoot them. Not all the prisoners were shot.

Q: Was Medina in the CP at the time this was taking place?

A: Yes.

Q: Did he know what was going on?

A: Yes.

Q: Was there ever an order given to stop the killings?

A: I did not receive the order but I assume it was given for the killings did stop.

Q: Did Co C meet any resistance in My Lai (4)?

A: No.

Q: Was any medical aid given at My Lai (4)?

A: No. The medics present was Cappezza, Lee, Fores, and Manzanedo. . . .

Q: Did you kill anyone in My Lai (4) on about 16 Mar 68?

A: No.

Q: Would you be willing to testify in court if you were called?

A: If I was called.

Q: Is there anything you would like to add or delete from this statement?

A: Only that I expected something to happen about that incident and I did not expect that it would wait this long.

30

LARRY POLSTON

Testimony to U.S. Army CID

1969

Larry Polston, a rifleman with the 2nd Squad of the 3rd Platoon, also refused to fire his weapon during the My Lai operation.

Q: What did you observe as you moved through the village?

A: ... We swept across the trail and at this time I noticed that there were 9–10 dead bodies scattered along this trail. These bodies were men, women and children of all ages. As I recall PFC Williams and PFC T'Souvas were with me. Someone in the Plt recovered the weapon from some VC and we returned to the rice paddy and regrouped outside My Lai (4). I did not see this VC but I do recall seeing one weapon but I can't recall who had it. I think it was an AK-47. While the Plt was moving to recover the weapon I had crossed the trail and as Williams, T'Souvas and I returned to the trail there were two small Vietnamese children laying down in the trail. They had already been wounded but I could see that they were alive. T'Souvas shot these children with the M-60 that he was carrying. Then a man and a woman dressed in civilian clothing came running down the trail away from My Lai (4). Williams shot and killed the woman and injured the man. We left him there along with the other dead Vietnamese. This made a total of 14–15 people shot along this trail. At this time I did not fire my weapon as I didn't see any sense in it. There was a lot of firing going around anyway. From here we returned to the rice paddy and regrouped and waited until CPT Medina gave orders to move into the village. I estimate that I was on the ground and in the rice paddy area for about 30 minutes, before we moved into the village. Before we entered the village there was a lot of shooting going on. I didn't see anybody else shot. I could hear the shooting and I figured it was from 1st and 2nd Plts. When we moved out into the village, I remained near the center of the village moving towards the west. I started two fires of empty hooches as I moved through the village. I saw a lot of dead bodies laying around

---

CID Deposition Files, My Lai Investigation, CID Statement, file no. 69-CID011-00014, U.S. Army Crimes Records Center, Fort Belvoir, Virginia, p. 204.

the village and I estimate 30–40 people. They were mostly women and children and a few men. They were all ages, ranging from babies to old men and women. I didn't see any weapons as I moved through the village and as far as I know no one shot at me. I didn't see any resistance from any of the Vietnamese. I didn't see any of the dead Vietnamese in the village shot and I guess they were shot earlier by members of the 1st and 2nd Plts. As I reached about half way through the village an order came down from CPT Medina to cease firing. At this point I had been in the village about half an hour. The shooting stopped and 3rd Plt continued to move through the village and set fires to all the buildings. I didn't see anyone killed or any shooting after the cease fire order was given. We moved on through the village and formed on the west side of the village. I estimate that we were in the village for approximately two hours. We had lunch and that afternoon swept a couple of more villages and spent the night in the field between My Lai (4) and My Lai (5). I can't recall what we did the next day or so or how many nights we stayed in the field before returning to LZ Dottie.

Q: At any time on this mission were you fired upon by the Vietnamese or did they offer any resistance?

A: No not that I know of.

Q: Did you shoot any of the Vietnamese or shoot at them?

A: No.

Q: Did you fire your weapon at all during this mission?

A: Not that I recall.

Q: From talking to other members of the unit, almost everybody was reported to be shooting at the people or animals. Why were you not shooting?

A: Well it was more or less other guys doing it and I just wasn't shooting. I didn't see any sense in it at all.

31

VARNADO SIMPSON

Testimony to U.S. Army CID

1969

Varnado Simpson, a rifleman with the 2nd Squad of the 3rd Platoon, participated in the massacre and, tortured by guilt, readily confessed his crimes to army investigators.

I, Vernado Simpson want to make the following statement under oath:

The next day we went to MyLai (4). I was in the second or third lifts. Another Platoon, the First Platoon of Company C, went in ahead of us. I was with my unit, the Third Squad, 2d Platoon. My Platoon Leader was Lt Brooks. My Squad Leader was Sgt LaCroix. My Platoon Leader or rather Platoon Sergeant was Sgt Buchanon. . . .

After we landed we advanced by fire into the village. We started on the left, but during the advance through the village the troops were all mixed up. Some of the 1st Platoon got with the 2d Platoon and so forth.

Just after we got into the village, I came upon Wood and Stanley with four or five Vietnamese detainees. Stanley said they were going to take them to the Platoon Collection area. They were asking these people some questions in Vietnamese. Then Roschevitz, who had come up with me, said to kill all the people and told me to kill them. I hadn't killed anyone yet, so I said that I would not. Then Roschevitz grabbed my M16 away from me and put it on automatic fire and killed all of the Vietnamese who had been standing there. These people were not armed and were not trying to escape.

Q: What happened then?

A: I continued on into the village and found a place where a boy had been shot by a well near a hut. A woman, carrying a baby, came out of the hut crying and carrying on. Roschevitz, Lamartina, and LaCroix were there. Wright, Hutto, and Hudson were there also. I think Brooks may have been around. Brooks told me to kill the woman, and, acting on his orders, I shot her and her baby. I have been shown a group of photographs and I identify the photograph of the woman and the baby

---

CID Deposition Files, My Lai Investigation, CID Statement, file no. 69-CID011-00069, U.S. Army Crimes Records Center, Fort Belvoir, Virginia, pp. 1–3.

as being the ones I shot as related here. I remember shooting the baby in the face. . . .

Q: What happened then?

A: There were four or five people—mostly children—still in the hut. Hutto, Wright, and Hudson went into the hut and Hudson fired the machinegun into the children. I had gone into the hut at that time and saw that the bodies were all torn up and I have no doubt they were all killed. There was a little old hole in the hut where the people took shelter from attack, and Wright dropped a grenade into the hole, in case someone was hiding there.

Q: What happened then?

A: As we moved into the village we heard a lot of firing and then came on an area where the platoon ahead of us had rounded up 25 or 30 people and executed them. We did not see the shooting, but it had just happened. Medina was there when we got there, but I don't know if he had witnessed the killing while it was going on. I heard about another execution that day not far from this scene (but didn't see it either during the killing of the people or afterwards), and also found a ditch full of people at MyLai (4).

Q: What happened next?

A: We were on the left, moving ahead and burning huts and killing people. I killed about 8 people that day. I shot a couple of old men who were running away. I also shot some women and children. I would shoot them as they ran out of huts or tried to hide.

Q: Did you see anyone else killed?

A: Yes. I saw Wright, Hutto, Hudson, Rucker (deceased), and Mower go into a hut and rape a 17 or 18 year old girl. I watched from the door. When they all got done, they all took their weapons, M-60, M16's, and caliber .45 pistols and fired into the girl until she was dead. Her face was just blown away and her brains were just everywhere. I didn't take part in the rape or the shooting.

Q: Did each of these men—Hutto, Wright, Hudson, Rucker, and Mower—have sexual intercourse with that girl?

A: Yes they did.

Q: Did each of them—Wright, Hutto, Hudson, Rucker, and Mower—fire into the girl after the act of intercourse was completed?

A: Yes they did.

Q: Did you see anyone else killed?

A: I witnessed a lot of people being killed, but there was a lot of confusion going on and I can't relate details of every killing I saw. I estimate there were 400 people killed in MyLai (4). I would like to stress that everyone was ordered by Medina to kill these people: that the killing was done on his orders.

Q: You said you saw a ditch full of people. Please tell me about that?

A: The First Platoon had been there and gone when we arrived. We saw an irrigation ditch with 30–40 dead Vietnamese in it. They had all been just killed. Some had been killed in the ditch and some had made it to the top of the ditch, but they were all dead. I don't know who did this by name, but it was the First Platoon.

32

HUGH THOMPSON JR.

Testimony to Peers Commission

1970

Hugh Thompson Jr., a helicopter pilot, witnessing the massacre from the air over My Lai, landed his chopper between Lieutenant William Calley and a group of Vietnamese civilians to prevent more slaughter.

Q: What happened when you put the chopper down?

A: ... When I saw the bodies in the ditch I came back around and saw that some of them were still alive. So I sat [the helicopter] down on the ground then and talked to — I'm pretty sure it was a sergeant, colored sergeant — and I told them there was women and kids over there that were wounded — could he help them or could they help them? And he made some remark to the effect that the only way he could help them was to kill them. And I thought he was joking. I didn't take him seriously. I said, "Why don't you see if you can help them," and I took off again. And as I took off my crew chief said that the guy was shooting into the ditch. As I turned around I could see a guy holding a weapon pointing towards the ditch.

　　... And after that we were still flying recon over the village. The village was smoking pretty good. You couldn't get right over it. And we came around somewhere to the east of the village, and I saw this bunker and either the crew chief or the gunner said that there was a bunch of kids in the bunker, and the Americans were approaching it. There was a little open area, field, shaped sort of like a horseshoe, so I set down in the middle of that horseshoe, got out of the aircraft and talked with this lieutenant, and told him that there was some women and kids in that bunker over there, and could he get them out. He said the only way to get them out was with a hand grenade. I told him to just hold your men right where they are and I'll get the kids out. And I walked over towards the bunker, motioned for them to come out, and they came out. But there was more than women and kids. There was a couple—one or two—old men in there. I'd say about two or three women and then some kids. I got back in the aircraft, didn't take off, just put my helmet on or just plugged my helmet up and I called Mr. Millians who was flying the low gun cover and told him what I had and asked him if he'd come in and get them out of this immediate area back into an area that had not so much firing going on. And he came in and picked up half of them.

Q: Was he able to land in the same horseshoe-shaped area?

A: No, he landed outside the horseshoe-shaped area. He landed behind me.

Q: Yes.

A: And I walked the people to him. He could only take about half of them, and flew them out going back to Highway 521. He flew them, I would say, back up to the vicinity of Hoa My . . . because there was a road going off 521 about where he let them out. He came back and got the rest of them and took them up there also. I followed them back. That's how I know where he landed.

Q: How many people did you pick up all told?

A: . . . Sixteen, I don't remember today. . . .

Q: Now, let me come back here again. Tell me a little more about your discussion with the lieutenant? Did you ever identify the lieutenant?

A: Yes, sir.

Q: Who was he?

A: Lieutenant Calley. . . .

Q: Could you tell us what happened as you best recollect, not from the newspapers, but from the time itself?

A: I just did, sir. I told him to stop his troops after he told me the only way he could get them out, and he stopped them. My crew chief and gunner were outside the aircraft also, and I walked across a rice paddy towards the tree line the bunker was in. I got, oh, I would say within

10 or 15 meters of the tree line and motioned for them to come out. As they came out, I gathered them in a little group, and I called for my low gunship and said: "I got some people down here. Can you come in and take them out for me and get them out of this area?"

Q: And that's all that happened? There were no other words or actions?

A: To the best of my knowledge today, sir, there was no words that I can recall between myself and the man who appeared to be the lieutenant.

Q: Any other actions taken?

A: The gunship came in. The one ship came in, took half of them out, went and dropped them off, and then came back and got the rest of them. And I didn't say anything else to the lieutenant to the best of my knowledge today, sir. . . .

Q: Was there any form of altercation or argument between you and Lieutenant Calley or anybody else there . . . ?

A: When I got out of the helicopter, I told my crew chief and gunner to make sure I was covered real close.

Q: From that, I take it, you expected—you were being covered real close. Were you inferring for protection against VC or protection against something that might have been done to you from the U.S. side?

A: I was worried about getting shot, sir, because when I walked over to where the women and children were if the enemy would have started shooting I would have been in a crossfire from the friendly troops because I was between where the enemy was supposed to have been and where our friendly troops were, sir.

Q: Were you afraid of getting shot by our own forces or by the enemy? What I'm trying to get is when you said this to your doorgunners, were they protecting, would they be covering you from the friendly or the enemy side? Or both?

A: They were covering me from both sides, sir. But I'm not saying they were covering me from our troops. Charlie [Vietcong troops] could have been behind our troops also, sir.

Q: Did Lieutenant Calley threaten you with his M-16 or any other way at this time?

A: No, sir.

Q: Did he point his M-16 at you?

A: No, sir, I didn't have any weapons pointed at me. He might have been standing with the—he didn't have it thrown over his shoulder. I mean, I'm sure he had it in his hand. But it wasn't trained on myself, sir.

Q: Was this PFC Colburn covering you with his M-60?

A: Yes, sir. Both my crew chief and my gunner both had M-60's.

33

ERNEST L. MEDINA

Testimony to U.S. Army CID

1969

Captain Ernest Medina recounted to the Peers investigators his actions related to removing weapons from dead Vietnamese. He would later be tried for, and acquitted of, killing the Vietnamese woman in this incident.

Q: And what did COL Henderson conclude from his investigation?
A: Well, could I tell you what happened, sir?
Q: Yes.
A: O.K. The gun ships were going, had been in the area, put suppressive fire on the LZ, and they had reported where the people that they had killed with weapons were located, and we were told to make sure that we got the weapons because in previous actions before in the same area when they had killed people with weapons somebody picked—women would pick it up and run with them and hide them, and we would not get the weapons, and we were instructed to make sure we picked up the weapons. One of the light helicopters—I think it was an OH-13, with the pilot and two machine gunners on each side—was flying throughout the area and marking locations of bodies with weapons and he had thrown smoke throughout the area. I had already sent some squads from the rear security element to police up these weapons and this one pilot had called and said, "I am hovering over a body with a weapon. Can you send somebody to get it?" I told him, "Roger, I will have somebody over there." Well, the call came back that, "Nobody is showing up at the location I have marked with smoke where there is a VC with a weapon."

I guess they were monitoring the talk at the task force. MAJ Calhoun came on the radio and said, "Damn it, Ernie"—I think that is what he said—"get somebody over to get the weapons." I said, "I have a squad coming back to get the weapons and go to the areas," and he said, "Get somebody over there now." I said, "O.K., I will take a team and go over there myself." So we went to where he had dropped smoke, that I could see the smoke, and the first place I came up to was

CID Deposition Files, My Lai Investigation, CID Statement, file no. 70-CID011-00013, U.S. Army Crimes Records Center, Fort Belvoir, Virginia, pp. 102–03.

a Vietnamese male or two Vietnamese males and one woman and they had been hit by artillery or something, sir, because they were mangled pretty badly. There was nothing there, so we went on and I am not exactly sure how far it was, where the chopper was, but he was hovering about 20 feet over this woman and over the VC with the weapon, and I don't know, that is all I know, that it was a VC with the weapon. I walked up to the body. I seen that it was a woman. As I walked up the chopper pulled back, moved back. I seen it was a woman and I seen it didn't have any weapon laying around it and she appeared to be dead. She wasn't moving. I didn't turn her over or anything like that. The chopper, they could have seen it was a woman with no weapon, so I turned around to walk away and as I turned around to walk away I caught a glimpse out of the corner of my eye of something in her hand underneath her and she started to move and the first thing that went through my mind was, "You dumb bastard, you are dead." She had got a hold of something. I spun around and fired two or three times (The witness illustrated by standing and moving his body.) and I killed her and he, I know, reported this, and this is what I think, that I know of that he had referred to MAJ Calhoun.

34

NGUYEN HIEU

Testimony to Peers Commission
1970

Nguyen Hieu was a twenty-three-year-old resident of My Lai on March 16, 1968. His recollection of the incident follows.

Q: What is your name?
A: Nguyen Hieu.
Q: How old are you?
A: Twenty-five years old.

Q: Are you a native of Tu Cung?

A: Yes. . . .

Q: . . . Were you in your house on the morning of 16 March 1968 when the Americans came?

A: Yes, I lived there in 1968.

Q: Were you there on the morning of 16 March 1968 when the Americans came?

A: Yes, I was there that morning.

Q: How many other members of your family were there with you in the house that morning?

A: Five.

Q: What did you do when you heard the artillery fire?

A: For the first time early in the morning I heard artillery come in here (indicating) and American helicopters come into here (indicating) on the west side of the village. They came here and they took us from the bunker.

Q: Was the bunker near your house?

A: Yes, right here (indicating).

Q: Did all the members of your family go in the bunker?

A: My mother stayed in the house. I and the children went to the bunker.

Q: How long did you stay in the bunker?

A: About 2 hours.

Q: Did the Americans come near the bunker?

A: Yes, they came into the bunker.

Q: They came into the bunker?

A: Yes.

Q: And did they make you come out of the bunker?

A: When the Americans came to my house my mother came out of house, and the Americans then raped my mother and they shot her.

Q: They shot and raped your mother?

A: Yes, shot and raped my mother. My sister ran out of the bunker and they shot my sister and two children. . . .

Q: How many Americans were there?

A: Two Americans.

Q: Were they Caucasians or Negroes?

A: I saw only one black and one yellow.

Q: One black and one yellow. No white?

A: I saw one black, one yellow, and another I don't know exactly.

Q: Which one raped your mother?

A: The black soldier. . . .

Q: What did the white soldier do while the Negro soldier was raping your mother?

A: After they shot my mother, the white soldier checked the house to see that everybody was dead and then he went out. . . .

 . . . And later, a second group of Americans came in to burn the house.

Q: Were you the only one that stayed in the bunker?

A: Yes, I stayed alone.

Q: And your sister went out of the bunker and was shot?

A: My sister went out to help my mother and was shot.

Q: Were they all shot right around your house or did they take them some place else and shoot them?

A: They were all shot in the house.

Q: After the soldiers that shot the people left, how long were you in the bunker before the other soldiers came that burned the house?

A: About 40 minutes.

Q: About 40 minutes?

A: Yes.

Q: Did you see the soldiers that burned the house?

A: No, I did not see the Americans that burned the house.

Q: Did they shoot any livestock? Any animals, chickens, pigs?

A: They killed two buffalo.

Q: What did you do after the soldiers left?

A: After the Americans left I buried my mother and sister.

Q: I am sorry that your family was killed like this. Thank you for coming here today to help us.

35

NGUYEN BAT

Testimony to Peers Commission
1970

Nguyen Bat was a forty-year-old resident of My Lai on March 16, 1968. His recollection of the incident follows.

Q: What is your name?
A: Nguyen Bat.
Q: How old are you?
A: Forty-two years old.
Q: Have you lived all of your life in Tu Cung?
A: Yes, I live in Xom Lang, within Tu Cung Hamlet.
Q: Were you present in Xom Lang on the morning of March 16th, 1968, when the Americans came?
A: Yes.
Q: Were you in your house at the time?
A: Yes. . . .
Q: What did you do when you first heard the artillery fire? . . .
A: When I saw the Americans coming, I went to the bunker.
Q: Did you hide in the bunker?
A: Yes.
Q: Was your family with you in the bunker?
A: No.
Q: Did the Americans make you come out of the bunker?
A: The Americans did not see me. . . .
Q: How many Americans did you see?
A: Eight.
Q: What were the Americans doing when you saw them?
A: When I saw the Americans they were coming and shooting people in the hamlet.
Q: Did they shoot people in your family?
A: Yes, in my family.
Q: How many people in your family?
A: Eleven.

Peers, *Report,* vol. 2, bk. 32, pp. 1–6.

Q: Did you see them shoot all 11 in your family?

A: Yes.

Q: What position were the Americans in when they shot the 11 members of your family? Will you please show on the picture?

A: The Americans came from this way (indicating . . . from the center of the village toward the northeast corner . . .).

Q: And then what happened? Where were the 11 members of your family when the Americans shot them?

(Witness indicated . . .)

Q: Did just one American shoot them, or did more than one American shoot them?

A: I saw eight Americans coming, but I do not know how many of them shot.

Q: Did you stay hidden at the point marked D until after the Americans had left?

A: Yes.

Q: Did you see any Americans shoot anyone else while you were hiding at point D?

A: I saw the Americans come to my house . . . and kill in my house and come to this house.

Q: Did you see the Americans take a group of people across these paddy fields toward this ditch over here?

A: I saw somebody dead here, but I did not see the Americans do it.

Q: Did you see a single helicopter come down and land in this area while you were hiding at point B?

A: No.

Q: How long did you stay hiding in the bunker?

A: About 5 hours.

Q: That's all, 5 hours?

A: Yes.

Q: What did you do after the Americans left?

A: After the Americans left, I came out of bunker, and came to see my family.

Q: Was your house burned?

A: Yes, burned. Two groups of Americans came to my house.

Q: Two groups?

A: Yes, two groups. The first group came to shoot, and the second group came to burn the house. . . .

My eight children were burned.

Q: Your eight children were burned?

A: Yes.

Q: Were your eight children shot before they were burned?
A: Yes, shot first, then burned. . . .
Q: I see. Thank you very much for coming to help us today.

36

Summary of Rapes

1970

Approximately twenty Vietnamese women were raped during the My Lai assault. The following document, compiled by army investigators, summarizes those rapes.

RAPE VICTIMS	SOURCE OF INFORMATION
1. Nguyet, Nguyen Thi, Age 15	Tha, Do Thi, CID, 29 Dec 69 "Unknown farmers told her that a girl Nguyen Thi Nguyet, 15, had been raped by American soldiers and then shot and killed."
2. Nguyet, Nguyen, Age 12	Hoa, Nguyen Thi, CID, 30 Dec 69 "She stated that she had heard that Miss Nguyen Nguyet, 12, was raped and then killed by Americans. She stated that Nguyet's mother, Mrs. Bu, told her this."
3. Nho, Pham Thi, Age 22	Quy (Qui), Truong, CID,
4. Man, Do Thi, Age 12	3 Jan 70
5. Muoi, Pham Thi, Age 11	"While going through the village he saw three dead girls

CID Deposition Files, My Lai Investigation, Vietnamese Statements, Rape Victims, U.S. Army Crimes Records Center, Fort Belvoir, Virginia, pp. 221–22.

at three separate houses, which he stated looked as though they had been raped. They were Pham Thi Nho, 22, Do Thi Man, 12 and Pham Thi Muoi, 11. All had been inside their homes, which were burned and all had been naked. Quy stated that the vagina of each girl had been ripped and they looked as if they had sexual intercourse." Nho shot in stomach.

6. Nho, Phong Thi, Age 18

7. Muoi, Phong Thi, Age 13

Ngu, Truong, CID, 4 Jan 70 "He heard that Phong Thi Nho, female, age 18, was raped and killed by the soldiers in her home and Phong Thi Muoi, age 13, was also raped. He did not know if the soldiers killed her."

8. Nguyet, Do Thi, Age 14

9. Nho, Pham Thi, Age 19

Khoa, Nguyen, CID, 25 Dec 70 "Khoa heard from someone that Do Thi Nguyet, age 14, was raped by some soldiers and later found dead. He did not know how she died, but was told she did not have any bullet wounds on her body when she was found. He also heard that Pham Thi Nho was raped and later shot by soldiers. He thinks she was about 19 years old."

10. Nho, Mrs. (NFI)

Do, Thiet, CID, 24 Dec 69 "He further heard that Mrs. Nho had been raped, shot and her body left in her burning house."

11. Nguyet, Do Thi

Tro, Pham Thi, CID, 27 Dec 69
"She did not witness any
women being raped or
molested, but did hear that
Do Thi Nguyet, age 14, was
raped by the American soldiers and later died."

12. Nho, Pham Thi, Age 28
13. Mui, Pham Thi, Age 13
14. Nguyet, Do Thi, Age 13

Tri, Nguyen, CID, 3 Jan 70
". . . Lieu's daughter Pham
Thi Nho, 28 (not married)
Pham Lieu's daughter Pham
Thi Mui, 13 and a girl that
stayed with Ba Lieu, Do Thi
Nguyet, 13. All were dead
shot by the Americans in
their house. Nguyet, Mui and
Nho were in Tri's opinion
raped before being shot. Tri
based this opinion on the fact
that the other bodies were
clothed while the three were
nude. Also the vaginas of
these three women were
ripped."

15. Nguyet, Do Thi, Age 10

Vien, Do, CID, 4 Jan 70
"Nguyet was raped by the
Americans. Vien based this
opinion on the fact that he
found the body in Ba Xam's
house and he saw clothes had
been torn off and her vagina
had been ripped and blood all
over that area. Examination
by him determined that there
were no bullet wounds in the
body. He could not see any
bruises on the body."

16. Hoa, Do Thi, Age 18
17. Nho, Do Thi, Age 26
18. Sam, Nguyen Thi, Age 45

Thu, Nguyen, CID, 28 Dec 69
"He heard from Nguyen Thi
Gan and Thi Ho, who now

live in Son My, that Do Thi Hoa, 18, Do Thi Nho, 26, and Nguyen Thi Sam, 45, had been found dead with no clothes on and the soldiers had raped them and shot all dead."

19. Nho, Pham Thi, Age 22

Co, Nguyen, CID, 15 Jan 70 "He heard that Pham Thi Nho, 22, and an unknown girl, 12, had been raped and killed."

20. Hoa, Tran Thi, Age 20

Ba, Co, CID, 23 Dec 69 "He stated that he did not see anyone raped but heard that Tran Thi Hoa, 20, was raped and now lives at an unknown address in Saigon."

37

DO VIEN

Testimony to U.S. Army CID

1970

Do Vien, a My Lai resident, was out in the fields when the American artillery barrage began. When American troops began landing in heli-copters, he fled the village and did not return until three in the afternoon. Vien summarized for the Peers investigators what happened to his friends and neighbors in My Lai village.

CID Deposition Files, My Lai Investigation, CID Statement, U.S. Army Crimes Records Center, Fort Belvoir, Virginia, pp. 207–10.

1. Truong Qui (Also known as (AKA) Truong Quy), age 40
   Pham Thi Mien, wife, age 35
   Four children but Vien did not recall their names.

Qui is presently civilian Detainee, LZ Bayonet; Mien and children are residing at Son My village.

2. Do Tuan, age 45
   Nguyon thi Em, age 40, wife
   Do Thi Nhan, age 16, daughter
   Do Dung, age 5 and one more child, name and age unknown.

Tuan is dead, stepped on a land mine sometime after the incident in question. Dung was slain in their house by the Americans. Em, Nhan and other child are living in Truong An refugee camp.

3. Do Phu, age 65
   Nguyen thi Cung, age 55, wife
   Do Tro, age 39, son
   Do Nhan, age 28
   Duong, age unknown, Nhan's wife
   Do thi Be, age 6, Nhan's daughter

Phu was killed near the well by the Americans, Tro was killed trying to escape by the Americans. The rest were slain in their house.

4. Do Vien, age 40
   Le thi Nguyen, age 28, wife
   Do Hat, age 6, son
   Do Ha, age 2, son
   Do Nguyen, age 60, Vien's father
   Pham thi Lat, age 55, Vien's mother
   Do thi Cang, age 23, Vien's sister
   Do Thanh, age 10, Vien's brother

Thanh, Cang and Ha were killed near the main trail approximately 30 meters from Vien's house. Vien is presently a civilian detainee, LZ Bayonet, RVN. The rest of his family reside at Truong An refugee camp, RVN.

5. House belonged to Do Dinh Luyen but was vacant when Americans came. Luyen is presently the Son My representative and lives in Quang Ngai City, RVN.

6. House belonged to Do Huan, age 60, Luyen's father and moved from the hamlet the same time as Luyen.

8. Do Thin, age 40, widower
   Do Thi Thuy, age 18, daughter
   Do Thi Co, age 70, Thin's mother

Thin and Co were slained at the irrigation ditch by the Americans. Thuy lives in Saigon. Thin and Co were cousins of Vien and he buried the bodies in the northeast section of the village.

9. Pham thi Nhi, age 65, widow
   Truong Ngu, age 37, son
   Pham thi Tri, age 32, Ngu's wife
   Truong thi Thu, age 10, Ngu's daughter
   Truong thi Le, age 32, Nhi's daughter
   Truong thi Bi, age 12, Le's daughter
   Truong Dung, age 6, Le's son

Nhi, Ngu, Tri, Thu and Bi were killed by the Americans at the dirt road leading into the village. Le and Dung are living at Truong An Refugee Camp, RVN.

10. Nguyen Gien, age 70, widower
    Nguyen Thi, age 40, son
    Pham thi Em, age 35, Thi's wife
    Thi's four children but Vien could not recall the names or ages.

Gien and two of Thi's children were killed on the road leading to the village. Thi, Em and the other two children live in Saigon, 3d precinct.

11. Le thi Ngu, age 65, widow
    Le Lien, age 30, member of ARVN and was not present when Americans came.

Ngu presently living at Son Tinh District refugee collection point.

12. Pham Thi Thia, age 60, widow
    Nguyen Co, age 35, son
    Do Thi Nhan, age 30, Co's wife
    Nguyen Cu, age 4, Co's son

Thia was killed by the Americans approximately 20 meters from the watchtower on the road leading to the village. Co was in the same group but escaped. Co, Nhan and Cu are presently residing in Truong An refugee camp, RVN. Co former VC that Chien Hoi'd [converted to U.S. side] Dec 69.

13. Nguyen Xam, age 60
    Nguyen thi Chua, age 60, wife
    Nguyen Mi, age 25, son
    Nguyen Xuan, age 23, son
    Nguyen thi Hoa, age 17, daughter
    Nguyen thi No, age 13, daughter
    Nguyen Cu, age 10, son

Chua and Cu were killed on the road leading to the village. Mi, Xuan, and Hoa are living in Saigon, 3d Precinct and Xam and No are living at Truong An Refugee Camp, RVN.

14. Nguyen Gap, age 50
    Trung thi Huyen, age 45, wife
    Three children, Vien could not recall their names or ages.

All were slain in their house and their bodies burned when the house was set afire by the Americans.

15. Nguyen Thai, age 45
    Pham thi Ri, age 40
    Nguyen thi Tam, age 15, daughter
    Nguyen Tu, age 10, son
    Nguyen thi Be, age 12, daughter
    Nguyen Go, age 6, son
    Nguyen Cu, age 2

All escaped and are presently living in Saigon, 1st Precinct

16. Nguyen Nhung, age 65
    Do Thi Hanh, age 60, wife
    Nguyen thi Nhung, age 32, daughter
    Nguyen Huy, age 20, son
    Le thi Han, age 18, Huy's wife
    Huy's two children, Vien could not recall the names or ages
    Nguyen thi Vang, age 20, Nhung's daughter
    Pham Chuong, age 30, Vang's husband
    Nguyen Cu, age 2, Chuong's son

Nguyen Nhung, Hanh, Huy, Han, Huy's two children, Vang and Cu were slain in their house by the Americans. Nguyen thi Nhung lives at Truong An refugee camp and Chuong is an ARVN, station unknown.

18. Do Nguyen, age 65
    Nguyen thi Xuc, age 60, wife
    Do Em, age 35, son
    Do Tui, age 32, son
    Do Mi, age 30, son
    Do thi Bay, age 25, daughter
    Pham Chim, age 30, lived in Binh Dong and was not present
    the day of incident.
    Bay's son, name and age unknown

Bay, Bay's son, Xuc and Nguyen were killed in their house by the Americans. Tui, Mi and Em are living in Saigon.

21. Truong Tho, age 80, widower
    Truong Cham, age 45
    Le thi Bay, age 40, Cham's wife
    Truong Khiet, age 15, Cham's son
    Truong Chan, age 17, Cham's son
    Two more children of Cham but names and ages unknown
    Truong thi Ri, age 40, Tho's daughter
    Truong thi Xuu, age 12, Tho's granddaughter

Tho was killed near the well and thrown into the well. Ri, Xuu, Bay and one unknown child were killed near the watchtower in the road by the Americans. Cham, Khiet, Chan, and the other unknown child are living in Truong An refugee camp, RVN.

22. Truong Long, age 65
    Do Thi Lien, age 60, wife
    Truong Van, age 32, son
    Do thi Loi, age 28, Van's wife
    Truong thi Ba, age 8, Van's daughter
    Truong thi Long, age 4, Van's daughter
    Truong Lien, age 18, Long's son
    Truong Thang, age 15, Long's son
    Truong Thang, age 10, Long's son

Do thi Lien, Ba, Loi, and Lang were slain in their house by Americans. All the rest are living in Truong An refugee camp.

23. Truong Xung, age 70
    Wife's name unknown
    Truong Truc, age 15, son

Xuan and wife were killed on the road near the watchtower by the Americans. Truc is alive but location unknown.

24. Nguyen thi Tri, age 40, widow
    Nguyen thi Tri, age 23, daughter
    Pham Phuong, age 30
    Cu, age 2 (NFI), Phuong son
    Nguyen Ky, age 6, Elder Tri's son
    Elder Tri's has two more children but Vien could not recall names or ages.
    Mother of Elder Tri, name and age unknown

Phuong, both Tris, N Cu, Ky and Elder Tri's mother were killed on the dirt road near watchtower by the Americans. The two unknown children were adopted by other villagers, names unknown are presently living Son Quang Village.

25. Nguyen Hoa, age 65
    Wife's name could not be recall[ed]
    Nguyen thi Hoa, daughter, age 23
    Do Chien, age 30, Hoa's husband
    Chien's two children, names and ages could not be recalled.
    Nguyen Dinh, age 16, Nguyen Hoa's son

Nguyen Hoa, his wife and one of Chien's children were killed in their yard by the Americans. Nguyen thi Hoa, her other child and Dinh are living in Truong An refugee camp. Chien is an active member of the National Liberation Front and was not in the area when the Americans came.

26. Pham thi Day, age 45, widow
    Do Duoc, age 6, son
    Do thi Nguyet, age 10
    Three more children that Vien could not recall the names.

Nguyet was raped by the Americans. Vien based this opinion on the fact that he found the body in Ba Xam's house and he saw clothes had been torn off and her vagina had been ripped and blood all over that area. Examination by him determined that there were no bullet wounds in the body. He could not see any bruises on the body. Duoc was killed in the vicinity of the well. Day and her three other children are living at Truong An refugee camp.

27. Pham thi Xe, age 75, widow
    Two nephews but Vien could not recall the names

All are living in Truong An refugee camp.

28. Pham Di, age 70, widower
    Pham Di, age 40, son
    Nguyen thi Niem, age 30, younger Di's wife
    Pham thi Pham, age 12, younger Di's daughter
    Pham thi Lang, age 10, younger Di's daughter
    Younger Di had two more children but Vien could not recall
    their names.

Younger Di was one of the crowd of villagers marched to the dirt road
by the Americans. There he was pulled from the crowd, forced to lay
down on the road and shot in the side by an American and left for dead.
He did not die. He recovered and is presently living with the rest of his
family in Quong Ngai City, RVN.

29. Le thi Em, age 35, widow
    Truong Tat, Em's son, age 10
    Truong Cu, age 12, son

All are living in Truong An refugee camp, RVN.

38

WILLIAM L. CALLEY

Testimony at Court-Martial
1970

*At his court-martial trial, Lieutenant William Calley's attorney asked him
a number of questions about the actual assault on My Lai. The attorney then
asked him how Captain Ernest Medina reacted to the fact that the platoon
had stopped to guard the civilians in the ditch. Calley responded as follows.*

William Calley Court-Martial Transcripts, National Archives Complex, College Park, Maryland, pp. 3812–18, 3821, 3832.

A: Well, after Captain Medina called me that time and I told him what was slowing me down, he told me to hurry up and get my people moving and get rid of the people I had there that were detaining me. . . .

Q: What was the substance of that conversation between you and Captain Medina?

A: He asked me why I was disobeying his orders.

Q: All right, was anything else said by him?

A: Well, I explained to him why—what was slowing me down and at that time he told me to "waste the Vietnamese and get my people out in line, out in the position they were supposed to be."

Q: Did you finish the substance of the conversation?

A: Yes, sir, basically.

Q: What did you do?

A: I yelled at Sergeant Bacon and told him to go on and stop searching the hootches and get your people moving right on out, not the hootches—the bunkers. I started over to Mitchell's location, and I came back out and Meadlo was still standing there with a group of Vietnamese, and I yelled at Meadlo and asked him—I told him if he couldn't move all those people to "get rid of them."

Q: Did he give you any reply to that, any audible reply that you heard?

A: He gave me a reply—I don't know what it was, sir.

Q: And then what did you do?

A: I continued over to Mitchell's location, sir.

Q: Did you hear any firing at that time?

A: There was firing, yes, sir.

Q: From where?

A: At Mitchell's location, back behind me, the gunships were still placing—they were still—I could hear heavy volumes of fire starting back up in the village. Apparently the third element was starting to come into the village.

Q: All right. Then what did you do?

A: I continued over to Sergeant Mitchell's location, sir.

Q: Did you fire into that group of people?

A: No, sir, I did not.

Q: All right, go on with relating what happened and what you did.

A: I went up and I told—talked to Sergeant Mitchell, told him to get his men together, get on the other side of the ditch and showed him where I wanted his machine gun emplaced and told him to link up his men with Sergeant Bacon, sir.

Q: Now during or approximately at that same time, was there an incident that occurred which you did some firing?

A: Yes, sir, there is.

Q: Now I am specifically referring to an individual. Was there an individual man involved somewhere around there or killed about that time?

A: No, sir, there wasn't.

Q: After you had this conversation up there with Mitchell, what happened and what did you do?

A: I heard a considerable amount of firing to my north, and I moved up along the edge of the ditch and around a hootch, and I broke out in the clearing and my men were—had a number of Vietnamese in the ditch and were firing upon them, sir.

Q: Now when you say your men, can you identify any of the men?

A: I spoke to Dursi and I spoke to Meadlo, sir.

Q: Was there anybody else there that you can identify by name?

A: No, sir. There was a few other troops but it was insignificant to me at that time.

Q: What's your best impression of how many there were at the ditch?

A: Four to five, sir.

Q: Two of whom you can specifically identify, Meadlo and Dursi?

A: Yes, sir. I spoke to those two.

Q: Now what did you do after you saw them shooting in the ditch?

A: Well, I fired into the ditch, also, sir. . . .

Q: All right. Then when it was up in there—after you moved up along side the ditch, that's when you started firing into the ditch, is that correct?

A: Yes, sir.

Q: After that incident, what did you do?

A: Well, I told my men to get on across the ditch and to get in position after I had fired into the ditch.

Q: Prior to that time, had you received any other message from Captain Medina?

A: Still to hurry up and get my men out in position where they were supposed to be.

Q: Was that the full extent of that conversation?

A: That's all I can remember him saying the whole time—from the time the second platoon broke and I got out in my position, sir.

Q: All right. After, did you leave the vicinity of the ditch—shortly thereafter, after you fired in the ditch, when did you leave it? Was it shortly—did you stay there a long time, or did you stay there a long while—a short time or what did you do?

A: I don't take it—it was a very rapid period of time to me. I can't say basically what time it was. It seemed like it was only a matter of half a minute, maybe a full minute at the most. I fired in the ditch and the men started moving across. Sergeant Bacon's men were coming out of the village, moving forward at that time. I started moving over to his location.

Q: Did you have a chance to look and observe what was in the ditch?

A: Yes, sir.

Q: What did you see?

A: Dead people, sir.

Q: All right. Did you see any appearance of anybody being alive in there?

A: No, sir.

Q: Let me ask you—at any time that you were along near that ditch, did you push or have anybody push people into the ditch?

A: Yes and no, sir.

Q: Give us the yes part first.

A: When I came out of the hedgerow, I came right out. The last man to go into the ditch—and I didn't physically touch him—if he would have stopped, I guess I would have.

Q: Was somebody there with him that was ordering him in or pushing him?

A: They had been ordered to go to the ditch, sir.

Q: Do you know who gave them that information?

A: Indirectly, I did, sir.

Q: Indirectly, what do you mean by that? Was it through somebody?

A: I had told Meadlo to get them people on the other side of the ditch, sir.

Q: After the incident where you went and you saw this one man and you fired into the ditch, what did you then do?

A: I continued moving northwardly along the edge of the ditch, sir.

Q: And for how far and where did you go?

A: Well, I went up to about the end of the ditch and about that time Warrant Officer Thompson, I believe he is Lieutenant Thompson now, landed his helicopter, sir. . . .

Q: All right. Now aside from what you said about the shooting into the ditch, was there any other shooting you did in that general vicinity?

A: The next time I fired, the helicopter had lifted off and I started walking over to the machine gun position and I fired on a head moving through the rice somewhere over in this area here (indicating). . . .

Q: There has been some information disclosed that you heard before the court that you stood there at the ditch for a considerable period of time, that you waited and had your troops organize groups of Vietnamese, throw them in the ditch or knock them in the ditch or pushed them in the ditch, and that you fired there for approximately an hour and a half as those groups were marched up. Did you participate in any such a shooting or any such an event?

A: No, sir, I did not.

Q: Did you at any time direct anybody to push people in the ditch?

A: Like I said, sir, I gave the order to take those people through the ditch and had also told Meadlo if he couldn't move them to "waste them" and I directly—other than that—it was only that one incident. I never stood up there for any period of time. My main mission was to get my men on the other side of that ditch and get in that defensive position and that's what I did, sir.

Q: Now why did you give Meadlo a message or the order that if he couldn't get rid of them to "waste them"?

A: Because that was my order. That was the order of the day, sir.

Q: Who gave you that order?

A: My commanding officer, sir, Captain Medina, sir.

Q: And stated in that posture, in substantially those words, how many times did you receive such an order from Captain Medina?

A: The night before in the company briefing, the platoon leaders' briefing, the following morning before we lifted off, and twice there in the village, sir. . . .

6

The Cover-Up

When they completed their investigation of the My Lai massacre, the Peers Commission concluded that a systematic cover-up of the incident had occurred at every level of command in the Americal Division. Warrant Officer Hugh Thompson, the helicopter pilot who had rescued some of the villagers, had reported the incident to his chaplain, Captain Carl Creswell, and to Colonel Oran K. Henderson, commander of the 11th Infantry Brigade, but a thorough investigation of the allegations never took place. It was not until Ronald Ridenhour's letter of March 29, 1969, more than a year after the massacre, that the army launched a formal investigation.

Since the end of World War II, when the United States helped codify the international laws of war, commonly Americans assumed that only ruthless dictators and their stooges committed war crimes. The United States military, it was widely believed, scrupulously observed the rules of war. But for several years before My Lai, antiwar activists had been accusing the Johnson administration of employing unnecessary force in South Vietnam, of trying to achieve a military solution to what was essentially a political problem, and of allowing too many South Vietnamese civilians deaths and casualties. By the time of the My Lai massacre, the antiwar movement had reached the summit of its power. At the end of March, President Lyndon B. Johnson announced that he would not seek reelection, hoping to focus his remaining time in office on bringing the Vietnam War under control and getting peace talks under way. The last thing his administration or the Pentagon needed in the spring of 1968 was a war crimes scandal. Such a political climate encouraged a cover-up.

So did military promotion policies. The Vietnam War had been a boon to career officers, giving them opportunities for the combat assignments they needed to achieve higher rank. Vietnam was a way of getting their "ticket punched." Lieutenants were anxious to impress their captains, captains their majors, majors their lieutenant colonels, lieutenant colonels their colonels, and colonels their generals. No officer wanted his personnel file full of information about war crimes committed under his command. Such a report would guarantee being passed over for promotion and could even end a promising career. Consequently, once the massacre had occurred, its magnitude was increasingly minimized at every level of command. Major General Samuel Koster was serving as commander of the Americal Division when the cover-up began. By the time the Peers inquiry was under way, he had been named commandant of West Point. Koster denied that there had been a cover-up of My Lai, insisted that a thorough investigation of the incident had been conducted, and claimed that he had seen a formal, written report. Peers investigators, however, were never able to locate the alleged document.

The moral implications of such a situation are ominous. If the slaughter of nearly five hundred Vietnamese civilians could be covered up for more than a year, in spite of dozens of eyewitnesses and a formal report to a full colonel who commanded an entire brigade, what other atrocities might have gone unnoticed during the Vietnam War? How many incidents of rape, murder, torture, and the unnecessary destruction of personal property had actually taken place over the years? Had the massacre been reported immediately, thoroughly investigated, and the perpetrators brought to justice, it might be easier to believe that My Lai was an isolated incident, an ugly aberration not typical of the war. But the systematic and temporarily successful cover-up of My Lai leaves the door open to another possibility: that the abuse of Vietnamese civilians was the rule, not the exception, of America's longest war.

39

HUGH THOMPSON JR.

Testimony to Peers Commission
1970

Determined to do all that he could to ensure that the My Lai incident was properly investigated, Warrant Officer Hugh Thompson met with Colonel Oran K. Henderson, commander of the 11th Infantry Brigade, the day after the My Lai attack. Peers investigators asked Thompson several times about his discussion with Henderson.

Q: When you talked to Colonel Henderson in the interview, can you give me your general impression of the interview? Whether you were telling him or whether you were answering questions, or whether he was distraught—the general nature of the interview?

A: He seemed interested, sir.

Q: Did he ask you what happened? Was he asking questions, or how did this all come out?

A: I think he asked me, you know, like what I had seen, and he asked me questions pertaining to a couple of things. You know, it's been a long time, sir, since I was in there, you know. It didn't strike a real great impression on my mind, sir, what I said back then, exactly, you know.

Q: You probably answered this before, but how long did you spend with him?

A: I think it was about 20 to 30 minutes, sir.

Q: You went over it pretty thoroughly with him, did you?

A: What I had seen.

Q: Did he ask you questions about the kind of dead people that you had seen?

A: I believe that I said that I didn't see but two or three draft-age males dead. But I can't just—it was a lot fresher on my mind about how many draft-age persons I saw, but, still, it was fresh on my mind so I feel sure that I probably would have mentioned that to him. But I can't swear that I did.

Q: Was he the only officer that interviewed you on the 17th of March?

A: Yes, sir. . . .

Peers, *Report*, vol. 2, bk. 8, pp. 43–46.

Q: ... Did you tell him about the situation with the captain on the ground and the other wounded around the area?

A: I know I told him about the captain. He said, "Do you know who he is?" and I said, "I don't know. How many captains did you have out there today?" I told him I didn't know the man's name, "But I don't know how many captains you had out there that day, but you probably didn't have a lot of them."

Q: Did you tell him about the other wounded that were near there that you had marked with smoke, and that you had seen? The ones you had marked with smoke and then called for help to provide medical assistance, and when you came back they were dead?

A: I can't swear to that, but I think I did.

Q: Yes. Did you tell him about the ditch with the bodies in it?

A: Yes, sir.

Q: Did you tell him about how many bodies you thought were in the ditch?

A: Yes, sir.

Q: Did you tell him that you had returned to the ditch to pick up the one boy at a later time?

A: Yes, sir.

Q: Did you also tell him about the bunker?

A: Yes, sir.

Q: Did you tell him about your conversation with the people near the bunker?

A: No, sir. I didn't tell him my conversation completely. I told him part of it.

Q: Yes. But did you tell him the fact that you had—that they had indicated to you that the only way to get them out was to grenade them out?

A: Yes, sir. I can't say that. I feel sure I did, but I can't remember my exact words to this man, sir. But I know that it upset me quite a bit that day. You know, it didn't upset me, it just kind of teed me off, I guess you'd put it. But I know good and well I'd mentioned it to him.

Q: Did you mention to him about your discussion with the sergeant at the ditch?

A: I feel that I did, sir.

Q: Yes. What frame of mind were you in when you talked to him? Did you consider yourself emotionally upset at this point of time? Distraught? Do you think you were clear in presenting the picture?

A: I think I presented it the best I could at that time. I'd say I was upset, or, you know, disturbed or something.

Q: Were you crying?

A: Oh, no. I wasn't crying, sir. ...

40

CARL CRESWELL

Testimony to Peers Commission
1970

When he returned from the My Lai operation on March 16, 1968, helicopter pilot Hugh Thompson was angry about the atrocities he had witnessed. Later that day, he met with Captain Carl Creswell, an army chaplain, and described the massacre. When Peers investigators talked with Creswell about the meeting, Creswell claimed to have taken Thompson's complaint to his immediate superior in the chaplain corps, Lieutenant Colonel Francis Lewis. As the following document indicates, the Peers investigators wanted to know why Creswell had not done more to see that a formal investigation was conducted.

Q: When did you . . . hear of the [My Lai] operation and under what circumstances?

A: It was, I'm willing to lay odds, on the 16th, the day of the operation—no later than the 17th—when Mr. Thompson came to see me. He said he had flown in, and he came in and sat down very upset. He was terribly upset and wanted to know what to do from that point on. It was my suggestion that he lodge an official protest in command channels, and I'd do the same thing through chaplain channels. I believe he saw General Young, and I saw Chaplain Lewis, the division chaplain. I told him about these allegations that had been made and that I had an awful lot of confidence in Mr. Thompson, and that I would—well, I'll be perfectly honest. I said that if there was not going to be an examination into these charges, I was going to resign my commission. . . .

. . . As I said, he [Thompson] was physically upset. I don't remember his exact words but he said it was a hell of a problem, and he didn't know exactly what he was going to do about it. He related the story of the day's operation to the effect that he had been assigned to lift some troops from Task Force Barker to Pinkville. And on a subsequent flight, he had seen an awful lot of bodies, from the air, of what appeared to be noncombatants. And he had landed at one point to evac civilians. He also told me that he had to threaten an American officer to lift them up and get them off the ground. After he left, they fired on the rest of the civilians and he was very, very angry. As a matter of fact, so was I because I believed him, and this sort of thing theoretically just doesn't happen. But theory and reality are two different things in Vietnam. . . .

Q: Once you had obtained this information, Father, what action did you take then?

A: I took it to the division chaplain and repeated the story in substance that Mr. Thompson had given to me. And I told him that Mr. Thompson was going through command channels, and I was going through technical channels, and I thought an investigation was most definitely called for because this sort of thing had to be proved or disproved. It can't be let lie around like a cancer. He assured me that he would take it higher, to the chief of staff, I suppose. I talked to him later, and he said that it had been in some manner brought to the attention of the chief of staff or division commander and that there would be an investigation.

Q: Did you submit your report to him in writing?

A: No, verbally.

Q: Did you at any time submit a report in writing?

A: No, sir.

Q: What happened after that, Father?

A: I would suppose—well, the article came out, of course, in *Stars and Stripes,* and from what I felt to be the facts of the case at that point, I thought it was ludicrous. I really did. Then the 11th Brigade paper came out with the story essentially the same as *Stars and Stripes.* As I recall, one of the things that struck me was the number 128 [killed] was terribly low compared with the number I got from Mr. Thompson. Once again, I can't even speculate on the number he gave me. I would suppose that it was about 2 weeks later that I asked Chaplain Lewis if the investigation was progressing, and he said that it was. Quite frankly, I dropped it there because, once again, the information I had was allegations and really had to be supported by Mr. Thompson's report through command channels and I'd be awfully glad to support him, but I couldn't carry the ball myself. It was really up to division to do something about it.

Q: Do you know who Chaplain Lewis talked to?

A: No, sir.

Q: Your superior?

A: I can only assume that it would have been the chief of staff, Colonel Parson, but that is an assumption.

Q: Quite a serious and, I would say, horrendous allegation had been given you. Do you consider in this instance that just reporting it to the chaplain was sufficient?

A: Yes, sir, for two reasons, the first of which is I knew that the same report was being forwarded through command channels.

Q: How did you know that?

A: Because Mr. Thompson told me that he had done this.

Q: But when nothing seemed to be happening about it, would it be appropriate to let an allegation such as this drop with the faith you had in Warrant Officer Thompson?

A: Well, sir, I was in a position where, really, all I knew was what he knew. His was the firsthand report. Mine was the secondary report. We went at it, really, from two directions. As I said, when it comes from Hugh Thompson to me, it's an allegation. When it comes from me to somebody else, it's hearsay. I really think that I went as far as I could. I was not first person singular there.

41

ERNEST L. MEDINA

Testimony to Peers Commission

1970

After his meeting with Hugh Thompson, Colonel Oran K. Henderson, commander of the Americal Division, met with Captain Ernest Medina to determine if there had been unnecessary civilian casualties at My Lai. He also inquired about allegations that Medina had killed a Vietnamese woman. Army investigators asked Medina several times about his meeting with Henderson.

Q: What were they [Henderson and Colonel Blackledge, the second-in-command of the 11th Infantry Brigade] investigating?

A: COL Henderson come out to the field and he talked to me and he asked me if I had killed this woman, if I had shot a woman, and I told him the only one I had shot was the one I told you about [see p. 93].

He asked me if I was aware of any civilians that had been killed—you know, murdered or killed by members of my company. I told him no. He asked me if I had given any such orders. I told him no. After I had explained about it, he said, "O.K., I understand that." He said, "That's understandable that something like that would happen." He said, "O.K., continue with your operation," and he left.

Then a couple of days later we were evacuated, picked up, and we started out with three ships. We had crossed back into Task Force Barker's AO, secured an LZ, and started lifting the company out. We started out with three helicopters and ended up with two, so it was rather slow.

When we got back I was informed that COL Henderson had been there and he had asked members of the company if they had seen any atrocities being committed and COL Barker told me about this also, so he asked me what I thought and I told him that as far as I knew none had been committed, that I had not ordered anything and I had not seen anything. He said, "O.K., the investigation is being conducted and, he said, "just forget about it and go about your own duties." . . .

Q: What day did COL Henderson first question you?

A: It was on the third day of the operation, sir.

Q: How did he question you?

A: Well, we were moving, conducting a search of the area. We were told there were some weapon caches and rice caches in the area and we managed to find some mines. I got a call on the radio saying that the brigade commander, COL Henderson, was coming to my location and to secure an LZ. I secured an LZ. He landed and got out of the chopper and the chopper took off and we went over to an area where we had some cover in case we received any fire and he just questioned me, sir.

Q: Did he give you an oath before you answered questions?

A: I don't believe so, sir.

Q: Did you sign any document concerning what you stated?

A: No, sir, I did not.

Q: Did he warn you of your rights before he questioned you?

A: I am not sure that he did, sir. No, sir.

Q: Do you know if this investigation or the investigation or inquiry later by COL Blackledge ever resulted in a written document?

A: Well, COL Blackledge, I don't know what part he played in the investigation, sir. I believe he was there with COL Henderson. For some reason I remember his face, but I don't know if he had anything to do with the investigation. If he did, they didn't tell me about it. COL Henderson did say he was conducting the investigation at the direction of the division commander. I think that was his exact statement, and that was it. . . .

Q: My question is did you place PFC Bernhardt on detail during the period this group of officers interviewed personnel of the company? Did you put Bernhardt on detail?

A: No, sir, Bernhardt was with me throughout the operation. I think he was with me throughout the entire operation and I think he went with me on the last ship. There was six of us left there on the LZ and we were the last ones out and this had already been conducted, so I believe he was with me.

Q: Did you tell Bernhardt at any time not to write his Congressman or give information on this operation?

A: Well, not exactly the way it sounds, sir. COL Barker had told me that COL Henderson had come around to check on this. He had talked to him. Bernhardt had written to Congressmen and the IG on numerous occasions before about small things and I believe it was LT Brooks who told me that he thought Bernhardt was going to write the IG or his Congressman and I called PFC Bernhardt over and tried to explain to him that there was an investigation being conducted and before he made any accusations or made, you know — something that was, may not even actually exist, to wait and see what the outcome of the investigation was, and I did not, the way it implies it, as you stated it to me, sir. . . .

Q: Was it [the possibility of atrocities] ever mentioned to you?

(Pause)

COUNSEL: I think we had best confer for just a moment.

Q: Yes.

(The witness and his counsel left the room for a short conference and upon their return the interrogation continued as follows.)

Q: I remind you you are still under oath.

A: (CPT Medina) Yes, sir. Sir, up until the time that the allegations were made, the investigation by COL Henderson, no, sir, I had no reason to suspect anything like this possibly could have happened.

Q: Well, what about subsequent to the investigation by COL Henderson?

A: You mean after that? Other than — it was — I don't know if it was COL Henderson or COL Barker that told me the investigation was being — well, when I come back COL Barker told me that COL Henderson had been there asking questions, if any atrocity had been committed, and I asked COL Barker why or what had happened and he said they had heard that a colored individual, a Negro, had been seen shooting into some bodies. They didn't know if they were dead or what. I asked him if I should make an investigation or start an investigation. They gave me a suspicion that something might have happened, but then I was told by COL Barker, "No, the investigation will be conducted and you just forget about it and go on about your duties."

Q: Was LT Calley's name ever mentioned?

A: Well, he was the 1st Platoon leader, but I can't say specifically—COL Barker or COL Henderson, no, sir, they did not.

Q: Then you did not investigate this allegation?

A: No, sir. I started to conduct an investigation of it and COL Barker said that investigation would be conducted and just to let it go.

TESTIMONY FROM CHARLIE COMPANY

Throughout 1969 and early 1970, investigators from the Peers Commission and the Criminal Investigation Division of the U.S. army interviewed members of Charlie Company. In addition to asking them about the assault itself, they tried to determine the thoroughness of the Americal Division's investigation of the incident. In the following depositions, soldiers from Charlie Company responded to questions about any investigation of My Lai.

42

MICHAEL BERNHARDT

Testimony to Peers Commission

1970

Michael Bernhardt was a rifleman with the 2nd Squad of the 2nd Platoon.

Q: Before you left the field, did anyone talk to you and tell you that there was an investigation going on of what happened at My Lai (4)?

A: I don't think so.

Q: Did Captain Medina get the company together later on and tell you that there was an investigation of what happened at My Lai (4) on 16 March?

A: He did that. I don't know what day it was. I thought it was when we got back to LZ Dottie.

Q: What do you recall about the meeting and what he said?

A: Just that there was an investigation and he would back us up and take our side for anything that happened. He gave the impression that he would accept responsibility or something like that, sort of reassuring them that it would be all right and there wouldn't be any problem.

Q: Did he say anything about the men should be quiet about what happened and not talk to anybody, or anything to that effect?

A: He told me.

Q: I know that. Now, how about the rest of the men?

A: I don't remember that, sir.

Q: What special instructions did he give you?

A: They called me up to the command post, and he said not to write my Congressman. I believe it was Sergeant Buchanon that told him. I don't know where he got the idea from. In other words, he didn't get the idea from anything I did or said, but it might just have been that he knew me or he thought he knew me. He told me it was not going to do any good, and it was going to get a lot of people in trouble and not to do it.

Q: Is this Captain Medina you are talking about?

A: Yes, sir.

Q: Did Sergeant Buchanon talk to you about it?

A: I don't think so, sir.

Q: Did he come back in the helicopter with you on the 18th?

A: I don't know whether he was in the helicopter with me or not.

Q: Did any investigating officer ever come around and talk to you about what happened at My Lai (4) on the 16th?

A: Nobody talked to me, sir.

Q: Did any investigating officer talk to any men in the company about it, about what happened at My Lai (4)?

A: I heard that there was a colonel that came down and talked to some of the men, and one of the men was Sergeant Buchanon.

Q: This was at Landing Zone Dottie when they came back?

A: Yes. I heard that he asked Sergeant Buchanon, "Do you think this is your job and this is what you are supposed to do?" And he said, "No comment," or words to that effect.

Q: Do you know whether any member of the company was questioned by an investigating officer about what happened at My Lai (4) on the 16th?

A: No, sir. I don't think so.

43

DENNIS CONTI

Testimony to Peers Commission

1970

Dennis Conti was a grenadier and mine sweeper with Lieutenant William Calley's 1st Platoon.

Q: Did you know Colonel Henderson?

A: No, I do not.

Q: He told us that he talked to several members of the company after they were lifted out, back to LZ Dottie, so this must have been the afternoon of the 18th. He assembled quite a few of them and talked to them, because he had, by this time, been told to look into what happened at My Lai (4). He said he asked them about any killing of noncombatant civilians. Nobody said anything much. He questioned two or three men but didn't learn anything. Were you present when he questioned any men?

A: I don't think so because I don't remember it. I was probably in a bunker trying to get some rest.

Q: Did anybody ever talk to you about an investigation?

A: Like I said, through the grapevine we heard that the warrant officer had gone back and complained to division, and we heard there was an investigation underway. That supposedly—you know, how supposedly everybody was supposed to be going to jail, and Captain Medina and all the officers were getting hung, and we were all going away for 150 years apiece. And then, later on, we were told that the investigation was dropped, and they told us we had a citation for it, because in the paper it read 128 VC killed.

Q: Could you put any times on these things you heard, for example, that the investigation had been dropped? How long was that, do you think, after My Lai (4)?

A: I don't know. Like I say, maybe a week or 10 days. I couldn't pinpoint it for you.

Q: Captain Medina told us that soon after the operation at My Lai (4) he called the company together, and told them that there was an investi-

Peers, *Report*, vol. 2, bk. 24, pp. 40–41.

gation going on, and that it might be better not to talk about what happened at My Lai (4) while the investigation was underway. Do you remember that?

A: No, I don't remember that either.

Q: Did anybody ever tell you to keep quiet about what happened at My Lai (4)?

A: I think there was something circulated through the company about that, too, about not saying anything. But like I said, the stuff you get it through the grapevine. Like I might not know, and somebody else tells me, and then I tell somebody else. I think I remember vaguely something being said. Whether it was said by an officer, an NCO, or a private, I don't remember. I know there was something said to that effect.

44

HERBERT L. CARTER

Testimony to Peers Commission

1970

Herbert Carter was a "tunnel rat" from the 1st Squad of Lieutenant William Calley's 1st Platoon.

Q: Did you ever hear anything about an investigation into the My Lai incident?

A: Yes.

Q: What did you hear?

A: I heard that they said if anybody asks around or any questions about what happened at My Lai, to tell them that we were fired upon and say that a sniper round had come in or something.

Q: Whom did you hear this from?

A: I was in the hospital at this time at Qui Nhon, and a couple of guys from the company came over. I'm not bragging, but most of the guys in that company liked me. I didn't bother nobody. I did my job and they did their job. We drank together.

Q: They came to see you in the hospital?

A: Yes. A lot of guys came over. You know, when they came back through, they would come over.

Q: Captain Medina told us that soon after this operation he got the company together and told them that there was an investigation and it would be better if nobody talked about it while the investigation was underway. Did your friends say anything about this?

A: No. The way they ran it down to me was like somebody was trying to cover something up or something, which I knew they were. They had to cover up something like that. . . .

Q: I think you know that it took a long time for the story of My Lai to get out. What is your opinion as to why this wasn't reported right at the time? You did mention about some of your friends coming and telling you to keep quiet. Do you know anything else?

A: Like a lot of people wondered how come I didn't say something. Now, who would believe me. I go up to you with a story like that and you would call me a nut. You would tell me I am a nut and that there was nothing like this going on. You would think that nothing like this goes on in the United States. Just like I was in a bar a couple of weeks ago, and there was a drunk in there. He was standing there reading a paper and he was asking me if I believed that things like that actually went on, and I said, "I wouldn't know, pal." It was kind of weird. This happened three different times. One time I was sitting up there with a friend of mine, and my partner told me to be quiet about the whole mess. Some people want to talk that talk all day long, and they just don't know this and that about what they are talking about.

45

ORAN K. HENDERSON

Report of Investigation

April 24, 1968

In late March 1968, Colonel Oran K. Henderson, commander of the 11th Infantry Brigade, was asked by the Americal Division Commander, Samuel Koster, to investigate and to determine whether a needless massacre of civilians had taken place at My Lai. Vietcong propagandists widely disseminated news of the massacre, and local Vietnamese officials in Quang Ngai province had complained about the incident. Henderson allegedly conducted an investigation and then sent the following report to the Americal Division.

Commanding General
Americal Division
APO SF 96374

1. (U) An investigation has been conducted of the allegations cited in Inclosure 1. The following are the results of this investigations.

2. (C) On the day in question, 16 March 1968, Co C 1st Bn 20th Inf and Co B 4th Bn 3rd Inf as part of Task Force Barker, 11th Inf Bde, conducted a combat air assault in the vicinity of My Lai Hamlet (Son My Village) in eastern Son Tinh District. This area has long been an enemy strong hold, and Task Force Barker had met heavy enemy opposition in this area on 12 and 23 February 1968. All persons living in this area are considered to be VC or VC sympathizers by the District Chief. Artillery and gunship preparatory fires were placed on the landing zones used by the two companies. Upon landing and during their advance on the enemy positions, the attacking forces were supported by gunships from the 174th Avn Co and Co B, 23rd Avn Bn. By 1500 hours all enemy resistance had ceased and the remaining enemy forces had withdrawn. The results of this operation were 128 VC soldiers KIA. During preparatory fires and the ground action by the attacking companies 20 non-combatants caught in the battle area were killed. US Forces suffered 2 KHA and 10 WHA by booby traps and 1 man slightly wounded in the foot by small arms fire. No US soldier was killed by sniper fire as was the alleged reason for

killing the civilians. Interviews with LTC Frank A. Barker, TF Commander; Maj Charles C. Calhoun, TF S3; CPT Ernest L. Medina, Co Co C,
1-20; and CPT Earl Michles, Co Co B, 4-3 revealed that at no time were
any civilians gathered together and killed by US soldiers. The civilian
habitants of the area began withdrawing to the southwest as soon as the
operation began and within the first hour and a half all visible civilians
had cleared the area of operations.

3. (C) The Son Tinh District Chief does not give the allegations any
importance and he pointed out that the two hamlets where the incidents
is alleged to have happened are in an area controlled by the VC since
1964. COL Toan, Cmdr 2d Arvn Div reported that the making of such allegations against US Forces is a common technique of the VC propaganda
machine. Inclosure 2 is a translation of an actual VC propaganda message
targeted at the ARVN soldier and urging him to shoot Americans. This
message was given to this headquarters by the CO, 2d ARVN Division
o/a 17 April 1968 as matter of information. It makes the same allegations
as made by the Son My Village Chief in addition to other claims of atrocities by American soldiers.

4. (C) It is concluded that 20 non-combatants were inadvertently
killed when caught in the area of preparatory fires and in the cross fires
of the US and VC forces on 16 March 1968. It is further concluded that
no civilians were gathered together and shot by US soldiers. The allegation that US Forces shot and killed 450–500 civilians is obviously a Viet
Cong propaganda move to discredit the United States in the eyes of the
Vietnamese people in general and the ARVN soldier in particular.

5. (C) It is recommended that a counter-propaganda campaign be
waged against the VC in eastern Son Tinh District.

> S/Oran K. Henderson
> T/ORAN K. HENDERSON
> COL, Infantry
> Commanding

46

ORAN K. HENDERSON

Statement to Peers Commission
1970

Because Colonel Oran K. Henderson's original report of his investigation claimed that there had been no wanton slaughter of civilians, army investigators asked him to describe in detail the process by which he had inquired into the massacre. Henderson prepared the following written statement explaining his role in the My Lai massacre and the subsequent investigation.

On the morning of 16 March I arrived in the objective area (approximately 0750 hours) after the artillery preparation and combat assault of C-1/20. . . . I observed the combat assault of Company B-4/3 in the Pinkville (My Lai 1) area and received a report that the initial assault was without enemy contact. . . .

While my aircraft was being refueled, the Division Commander (Major General Koster) arrived at LZ DOTTIE and we discussed the operation. The time was approximately 0930 hours. I informed General Koster that I had observed the bodies of approximately 6–8 civilians and from the position of their bodies, it appeared they had been killed by artillery fire. . . . I recall General Koster commented on the 6–8 civilians I had reported killed, and I believe he directed that I attempt to ascertain more accurately how they had been killed. . . .

Upon my return to LZ BRONCO that evening I telephoned the Division Commander and updated my report of the number of civilians killed. General Koster expressed concern over this report and directed that I have LTC Barker provide a breakout on these showing men, women and children and how they were killed. Although I had earlier placed this requirement on LTC Barker, I telephoned him again that evening to make him aware of the Division Commander's personal interest.

On the morning of 17 March I returned to LZ DOTTIE. At this time, LTC Barker provided me a 3″ × 5″ card which reflected the civilian casualties and how they had been killed. At this time, I was introduced to a Major Wilson, Executive Officer, Co. B, 123d Aviation Battalion. This major informed me that he had a Warrant Officer with him who had been on yesterday's operation and he desired that I hear the story direct from the Warrant Officer. I used LTC Barker's quarters and WO Thompson relayed to me generally the following information:

That the ground operation phase he observed appeared out of control. That both men on the ground and gunships were shooting at everything that moved. That he had seen what appeared to him to be civilian bodies all over the area. That he had seen a captain shoot a wounded woman. That he had marked the position of wounded civilians with smoke grenades and that small Infantry elements would then move to the area advancing by fire and movement. That he was not in radio communications with the ground elements. That he did not believe these actions were dictated by the tactical situation.

I informed WO Thompson that I would look into this matter and that if any disciplinary action was required, it would be taken. I also informed Major Wilson, Executive Officer, Co. B, 123d Aviation Battalion, that I recommend he report this incident to the Division Aviation Officer. Major Wilson had also informed me that none of his other pilots had reported observing anything similar to WO Thompson's report.

I departed immediately to the field location (vicinity of My Lai 1) of C-1/20 and met with Captain Medina. I informed him of the report made by WO Thompson, Co. B, 123d Avn Bn, and asked if he was the captain who had killed the wounded woman. Captain Medina's response was immediate and direct. He stated that an OH 23 helicopter was hovering approximately 150 meters from his position and had twice dropped colored smoke signals which indicated to him enemy dead and armed. Captain Medina reported he had no troops in his immediate vicinity so following the second signal he moved to personally investigate. He observed on the ground a woman in her early twenties, and after he nudged her with his foot or weapon, and he got no response, he assumed she was dead. As he moved away approximately 10 steps, he caught a movement out of the corner of his eye and on instinct he whirled and fired his M16 into the body. Captain Medina claimed that at the moment he thought the woman was throwing a hand grenade. Captain Medina also reported that on no previous operation had smoke signals been used to identify civilian casualties, but always used to show armed enemy. Captain Medina assured me that neither he nor any of his soldiers had knowingly caused any civilian deaths.

Returning to LZ DOTTIE I met with Brigadier General George Young, Assistant Division Commander, and relayed to him the report received from WO Thompson and response of Captain Medina. Following the departure of Brigadier General Young, I directed LTC Barker to have Captain Medina's Company (C-1/20) . . . sweep back through the area to identify more specifically how the 20 civilians were killed, to determine the accuracy of this number and to search for hidden weapons. LTC Barker informed me that the extraction of C-1/20 was already scheduled, aircraft arranged, and that time was insufficient to accomplish a thorough search. I informed LTC Barker that if we could not reschedule, the extraction due to nonavailability of aircraft, we would walk the company out.

Late in the afternoon I learned that General Koster had visited LZ DOTTIE and he had directed that the extraction take place as scheduled. Upon receiving this information, I returned to LZ DOTTIE and arrived concurrently with two HUID's and one CH 47 load of troops from C-1/20. I assembled some 30–40 troops and spoke to them briefly.

I recall informing them that I appreciated the fine job they had done and it would be some time before the 48th Local Force Battalion could again mount an operation against us. I also informed them that some non-combatants had been reportedly killed and that this destroyed much of the mission's success. I also informed them that some of these noncombatants may have been killed without provocation and by members of their company. I then asked if anyone had observed anyone shooting unarmed civilians or any acts of a questionable nature. To this I received no response. I then directed the question to three or four individuals and in each case received a "No, Sir."

These soldiers appeared to be in high spirits and I observed them move down the trail. No individual made any effort to make contact with me.

On or about 20 March, I reported to General Koster in his office at Chu Lai and provided him orally a report of my inquiry into this incident. My oral report included the general statement made by WO Thompson, Co. B, 123d Avn Bn, and that except for the acknowledgement by Captain Medina, his report of observations could not be substantiated by either members of C-1/20, TF BARKER Headquarters personnel, nor pilots of the 74th Aviation Company. These latter personnel were interviewed by Major Gibson, Company Commander, at my request. . . .

On or about 20 April 1968 I received instructions from Brigadier General Young to prepare a written report regarding certain aspects of the alleged My Lai incident. . . . In preparing this report, I used previously held notes and followed this up with some informal discussions with members of the Brigade whom I do not now recall. I personally delivered my written report to Colonel Nels Parson, C/S, 23d Infantry (Americal) Division on or about 24 April 1968. . . .

During May 1969, I was requested to appear at OTIG [Inspector General's Office], Washington, D.C. to answer questions regarding the My Lai (4) incident. The investigation was being conducted by a COL Wilson. I was surprised to learn that copies of the reports cited above were not available. COL Wilson requested I attempt to secure these reports and provide them to him. Upon my return to Hawaii, I telephoned the Chief of Staff, American Division (COL Donaldson) and requested he have someone look into the 11th Bde S2 safe and send me a copy of my report. COL Donaldson informed me a few days later that the report had been found and forwarded to Hq, USARV, with a copy to me. Upon receipt of this copy, I forwarded it to COL Wilson, OTIG.

Except for the report from the WO pilot, Warlord unit, my informal query developed nothing to indicate any disciplinary action or to warrant further investigation. I discussed the killing of the 24 civilians in my Bde staff meetings as an example. Thereafter, I assured that every detail concerning civilian control and evacuation was included in operation plans and orders. It was not treated lightly at the time nor in subsequent operations.

At the time of this incident the 11th Infantry Brigade was operating with some 60 officer and enlisted positions short. These positions were being used by TF Barker. From my assumption of command on 15 March until late June I operated without a brigade executive officer. On 23 March 1968, I was wounded in the leg and for the next three to four weeks wore a cast and used crutches or a cane. It was also during March that the American Division infusion program commenced its impact within the brigade. Simultaneous, the R&R program commenced in March. I cite the above, not as an excuse but to reflect on conditions that could possibly have contributed to my inability to ferret out what now appears to be facts existing at the time. I was not a party to nor do I know of any collusion to deceive or play down any facts in this case. The impetus for the initial query into this alleged incident was generated by the unidentified WO pilot from the Warlord unit. I took the most readily means available and conducted what I considered an appropriate command inquiry. Consequently, I assume full responsibility for any errors or omissions.

ORAN K. HENDERSON
Colonel, Infantry

47

Vietcong Leaflet
1968

Soon after the massacre at My Lai, Vietcong cadres began to disseminate the story of the assault among the South Vietnamese civilian population. For years, Ho Chi Minh had told his followers that the war in Vietnam was primarily political, that the real war was to win the loyalties of South Vietnamese peasants. The My Lai massacre supplied the Communists with a powerful propaganda tool, and they used it to their advantage. Within days of the massacre, Vietcong printers published leaflets describing what had happened at My Lai and distributed them by the thousands to the civilian population. American military officials were soon compelled to determine whether the leaflets were accurate or simply shrill propaganda. The following document is typical of Vietcong efforts to take political advantage of the massacre.

The Americans Devils Devulge [*sic*] the Truth

The empire building Americans invade South Vietnam with war. They say that they came to Vietnam to help the Vietnamese people and that they are our friends.

When the US Soldiers first arrived in Vietnam they tried to conceal their cruel invasion. They gave orders to the US soldiers to be good to the Vietnamese people thus employing psychological warfare. They also employed strict discipline which required US soldiers to respect the Vietnamese women and the customs of the Vietnamese people.

When the first US soldiers arrived in Vietnam they were good soldiers and they paid when they made purchases from the people. They would even pay a price in excess of the cost. When they did wrong they gave money to indemnify their deeds. They gave the people around their base-camps and in nearby hamlets medical aid. US newspapers often printed pictures of US troops embracing the Vietnamese people and giving candy to children. The American Red Cross also gave medical attention to the Vietnamese. This lead [*sic*] a small group of ARVN's to believe that the American man was a good friend and had continued pity for the people. The Army Republic of Vietnam was happy to have allies which are such good friends and who are rich.

But, it is a play and every play must come to an end and the curtain

---

Peers, *Report,* vol. 4, pp. 264–65.

come down. The espionage was very professional and clever. If the plan is completed it will one day become saucy, because all the people will know what they are trying to hide and what they are really doing to the Vietnamese people.

They continue to produce this play but each year they receive fewer victorious responses. Each year they are attacked by the enemy in the south and they are being defeated more every day. This play lies to the people and will soon be disclosed to them. Today the Americans cannot cover anything. Now they only kill and rape day after day. Their animalistic character has been uncovered even by the American civilians. In Saigon there are some Americans that put their penis outside of their pants and put a dollar on it to pay the girls who sell themselves. The Americans get laid in every public place. This beast in the street is not afraid of the presence of the people.

In the American basecamps when they check the people they take their money, rings, watches, and the women's ear rings. The Americans know the difference between good gold and cheap bronze. If the jewelry is of bronze they do not take it.

Since the Americans heavy loss in the spring they have become like wounded animals that are crazy and cruel. They bomb places where many people live, places which are not good choices for bombings, such as the cities within the provinces, especially in Hue, Saigon, and Ben Tre. In Hue the US newspapers reported that 70% of the homes were destroyed and 10,000 people killed or left homeless. The newspapers and radios of Europe also tell of the killing of the South Vietnamese people by the Americans. The English tell of the action where the Americans are bombing the cities of South Vietnam. The Americans will be sentenced first by the Public in Saigon. It is there where the people will lose sentiment for them because they bomb the people and all people will soon be against them. The world public objects to this bombing including the American public and that of its Allies. The American often shuts his eye and closes his ear and continues his crime.

In the operation of 15 March 1968 in Son Tinh District the American enemies went crazy. They used machine guns and every other kind of weapon to kill 500 people who had empty hands, in Tinh Khe (Son My) Village (Son Tinh District, Quang Ngai Province). There were many pregnant women some of which were only a few days from childbirth. The Americans would shoot everybody they saw. They killed people and cows, burned homes. There were some families in which all members were killed.

When the red evil Americans remove their prayer shirts they appear as barbaric men.

When the American wolves remove their sheepskin their sharp meat-eating teeth show. They drink our peoples blood with animal sentimentality.

Our people must choose one way to beat them until they are dead, and stop wriggling.

For the ARVN officer and soldier, by now you have seen the face of the real American. How many times have they left you alone to defend against the National Liberation Front? They do not fire artillery or mortars to help you even when you are near them. They often bomb the bodies of ARVN soldiers. They also fire artillery on the tactical elements of the ARVN soldiers.

The location of the ARVN soldier is the American target. If someone does not believe this he may examine the 39th Ranger Battalion when it was sent to Khe Sanh where its basecamp was placed between the Americans and the Liberation soldiers. They were willing to allow this battalion to die for them. This activity was not armed toward helping South Vietnam . . . but was to protect the 6,000 Americans that live in Khe Sanh.

Can you accept these criminal friends who slaughter our people and turn Vietnam into red blood like that which runs in our veins?

What are you waiting for and why do not you use your US Rifles to shoot the Americans in the head—for our people, to help our country and your life too?

There is no time better than now
The American Rifle is in your hands
You must take aim at the Americans head and
Pull the trigger

48

National Liberation Front Committee Notice

March 28, 1968

The National Liberation Front Committee of Quang Ngai province began disseminating the following document shortly after the My Lai massacre.

Notice

CONCERNING THE CRIMES COMMITTED BY US IMPERIALISTS AND
THEIR LACKEYS
WHO KILLED MORE THAN 500 CIVILIANS OF TINH KHE (V), SON TINH (D)

The morning of 16 March 1968 was a quiet morning, just like every other morning, with the people of Tinh Khe Village about to start another laborious day of production and struggle. Suddenly, artillery rounds began pouring in from Nui Ram Mountain, Binh Lien and Quang Ngai Sub-Sector. Xom Lang Sub-Hamlet of Tu Cung Hamlet and Xom Go Sub-Hamlet of Co Luy Hamlet were pounded by artillery for hours. After the shelling, nine helicopters landed troops who besieged the two small sub-hamlets. The U S soldiers were like wild animals, charging violently into the hamlets, killing and destroying. They formed themselves into three groups: one group was in charge of killing civilians, one group burned huts and the third group destroyed vegetation and trees and killed animals. These American troops belonged to the 3d Brigade of the [Americal] Division which had just come to Viet Nam and suffered a defeat in the Spring. Wherever they went, civilians were killed, houses and vegetation were destroyed and cows, buffalo, chickens and ducks were also killed.

They even killed old people and children; pregnant women were raped and killed. This was by far the most barbaric killing in human history.

At Xom Lang Sub-Hamlet of Tu Cung Hamlet, they routed all the civilians out of their bunkers and herded them, at bayonet point, into a group near a ditch in front of Mr. Nhieu's gate (Mr. Nhieu was 46 years old). About 100 civilians who squatted in a single line were killed instantly by bursts of automatic rifle fire and M79 rounds. Bodies were sprawled about, blood was all over. Among those killed were 60 year old men and newly born babies still in their mother's arms. Most of them were children from 1 to 14 years old. The wounded children who were screaming were shot to death.

Peers, *Report*, vol. 4, exhibit M-35, pp. 145–48.

Some entire families were massacred. Inhabitants were killed inside bunkers, in the gardens of their homes or in the alleys of the hamlet. Mr. Huong Tho, 72 years old, was beaten, his beard was cut, and he was pushed into a well and shot with automatic rifle fire until his body submerged. Nguyet, 12 years old, after being raped, was bayoneted in the vagina and rest of her body. Phan Thi Mui, 15 years old, was raped and then burned to death in a rice bowl. The entire seven members of Mr. Le Ly's family were killed, including the youngest 4 year old nephew and 70 year old Mr. Ly. The only survivor of the family was a married daughter who lived somewhere else and returned to the hamlet after the massacre, to cry, holding her beloved relatives' bodies in her arms. (She set up an altar at her father's burned house and prayed for her seven dead relatives.) Neighbors set up altars for families that had no surviving relatives.

The total number of civilians killed at Xom Lang Sub-Hamlet was 2060, including old people, children, women and young people.

At Xom Go Sub-Hamlet of Co Luy Hamlets, American pirates blew up and burned every hut and tossed grenades into civilian shelters. The sand was soaked with blood; beheaded bodies lay sprawled on the ground. People died without enough time to utter a word! Mothers holding sons' bodies! Grandmothers holding grandsons' bodies. They died unjustly. Fifteen people were killed inside Mr. Le's shelter. They even killed pregnant women. Vo Thi Hai, who had given birth to a child the night before, was raped and killed, leaving behind a newly born baby with no milk, with no one to suckle it. Nguyen Thi Ngon, 32 years old, near the end of her pregnancy, was mutilated inside her bunker, exposing the stirring, unborn baby. While 30 year old Vo Thi Phu was feeding her baby, they snatched her baby away and raped her. Later, both were burned to death.

Mrs. Kheo, 65 years old, was shot to death by the bunker entrance and her body was tossed onto the burning fire. Mr. Duong, 85 years old, was marched out of the bunker when they came. They marched him to every bunker, showing him the sights of the barbaric killings. They offered him poisoned candy, but he caught the bad smell and didn't eat the candy. They searched him and found nothing and released him.

At this place, American pirates killed 92, wounded 10, burned 304 huts, destroyed 78 bunkers and destroyed and burned civilian property worth 900,000 piastres.

Civilian laborers who had come to work or to visit relatives at Tu Cung Hamlet and Co Luy Hamlet were also massacred.

Thus, on 16 March 1968, the U S pirates and their lackeys massacred a total of 502 people at Tu Cung Hamlet and Co Luy Hamlet of Tinh Khe

Village and wounded 50 who survived the first bursts of automatic rifle fire. Among the dead were 67 old people, 170 children, 137 women. All huts, trees and animals were completely burned and destroyed.

This is by far the most typical of the barbaric massacres committed by the U S Imperialists against our People.

Like the other earlier massacres by American and Korean pirates at Phuong Dinh, Son Tinh; Van Ha, (Mo Duc) Phu Tho, (Nghia Hanh) Binh Hoa, (Binh Son) Pho Minh, (Duc Pho) the Tinh Khe Village massacre was the worst crime committed by the U S Imperialists and their lackeys before their complete defeat.

The Heavens will not tolerate this! The blue ocean waters will not wash away the hatred. These murders are even more savage than Hitler [or] Tan Thuy Hoan. [Translator's Note: Tan Thuy Hoan was the Chinese emperor who ordered hundreds of thousands of civilians to build the Great Wall, which resulted in thousands of deaths.]

Shamefully defeated, confused, the enemy is like a wild animal just before dying, due to our thunderous Spring attack. They have become excited and crazy, hoping to shake our spirit and the heroic tradition of our people. Quang Ngai is the province of proud Ba To, uprising Tra Bong, Ba Gia, Van Tuong, the province of glorious victories that scared the enemy.

With deep hatred in their hearts, the people of Tinh Khe, as well as the people and armed forces of Quang Ngai Province, have turned their sufferings and hatred into a rising, vengeful force.

After the massacre, the people of Tinh Khe Village wiped away their tears, hate deep in their hearts, and bravely rebuilt their homes, clearing away all traces of tragedy, growing potatoes, rice.

Immediately after the massacre, Tinh Khe guerrillas and other village guerrillas killed 31 enemy soldiers on 17 and 19 March 1968, including 17 Americans. Tinh Khe guerrillas personally killed 8 Americans on 17 March. The armed forces of the province have forced the enemy to pay their bloody debt.

During the 15 days from 13 March to 28 March 68, Local Force and Quang Ngai guerrillas fought many battles, killing 298 enemy soldiers, including 20 Americans, and captured much equipment.

The massacre of the 500 civilians of Tinh Khe has increased our hatred. We must attack continuously, rising up to make the enemy pay their debts!

National Liberation Front Committee
of Quang Ngai Province

Summary of Son My Chief's Letter
April 11, 1968

Early in April 1968, the village chief of Son My wrote a letter to Lieutenant Tran Ngoc Tan, the Son Tinh district chief, describing the slaughter of more than 450 civilians at My Lai. In the following document an American investigator summarizes the contents of the letter and the initial reaction by South Vietnamese district officials.

This statement is in reference to letter from the Son Tinh District Chief to the Quang Ngai Province Chief Subject: Allied forces Gathered People of Son-My Village for Killing, dated 11 April 1968.

The Son Tinh District Chief received a letter from the Village Chief of Son-My Village containing the complaint of the killing of 450 civilians including children and women by American troops. The Village Chief alleged that an American unit operating in the area on 16 March 1968 gathered and killed these civilians with their own personal weapons. The incident took place in the Hamlets of Tu-Cong and Co-Luy located in the eastern portion of Son Tinh District. According to the Village Chief the American unit gathered 400 civilians in Tu-Cong hamlet and killed them. Then moved to Co-Luy hamlet. At this location the unit gathered 90 more civilians and killed them.

The Son-My Village Chief feels that this action was taken in revenge for an American soldier killed by sniper fire in the village.

The letter was not given much importance by the District Chief but it was sent to the Quang Ngai Province Chief. Later the Son Tinh District Chief was called and directed by the 2d Division Commander, Col Toan, to investigate the incident and prepare a report. The District Chief proceeded to interview the Son-My Village Chief and got the same information that I have discussed above. The District Chief is not certain of the information received and he has to depend on the word of the Village Chief and other people living in the area.

The two hamlets where the incident is alleged to happen are in a VC controlled area since 1964.

Peers, *Report,* vol. 3, p. 263.

TRAN NGOC TAN

Report to Quang Ngai Province Chief
March 28, 1968

Lieutenant Tran Ngoc Tan, the Son Tinh district chief, was asked by South Vietnamese government officials in Quang Ngai province to investigate Vietcong reports that Americans had massacred civilians at My Lai. The following is Tran's report to the Quang Ngai province chief.

Son Tinh, 28 March 1968

Quang Ngai Province
Son Tinh District

No. 181/HC/ST/M

FROM: 1LT Tran Ngoc Tan
                     Son Tinh District Chief

TO: Quang Ngai Province Chief

SUBJECT: Confirmation of Allied Troops Shooting at the Residents of Tu Cung Hamlet, Coordinates BS 721795

It is respectfully reported that:

On 19 [sic] March 1968, an element of the US Forces (unspecified, because this District Headquarters had not been notified of the operation) conducted an operation at Tu Cung Hamlet (BS 721795), Son My Village, Son Tinh District. It was reported that when the element entered the hamlet, one of its members was killed and some others wounded by a VC booby-trapped mine. At this time the VC opened up fiercely from their positions in the hamlet. Meanwhile, the US troops used intense firepower while moving in with artillery and air support, inflicting injuries on a number of hamlet residents because the VC mingled with the population.

Observation by this Headquarters:

The Tu Cung Hamlet and the two neighboring hamlets, e.g., My Lai (BS 737800) and Van Thien (BS 794804), in Son My Village had become insecure since 1964, so the administrative authorities of these areas had

---

Peers, *Report,* vol. 3, exhibit M-5, p 15.

been forced to flee to Son Long (BS 638756), leaving these hamlets under VC control. Casualties were unavoidably caused to the hamlets residents during the firefight, while the local administrative authorities were not present in the area. The enemy may take advantage of this incident to undermine, through fallacious propaganda, the prestige of the RVNAF, and frustrate the Government's rural pacification efforts.

Respectfully yours,

(Signed and Sealed)

Copies to:
S2 and S3, Quang Ngai Sector HQ

51

THOMAS R. PARTSCH

Journal Entries
March 16–18, 1968

Several members of Charlie Company kept personal journals during their tour of duty in Vietnam, and many of them wrote about what happened at My Lai. Others wrote letters home to loved ones and described the massacre. Their journal entries and letters, written soon after the event, constitute primary evidence of an atrocity.

During a short break to drink some water during the My Lai operation, Thomas Partsch recorded the following entry in his journal. Two days later, on March 18, he wrote again, this time about an impending investigation.

<u>Mar 16 Sat.</u> got up at 5:30 left at 7:15 we had 9 choppers. 2 lifts first landed had mortar team with us. We started to move slowly through the village shooting everthing in sight children men and women and animals. Some was sickening. There legs were shot off and they were still moving it was just hanging there. I think there bodies were made of rubber.

Peers, *Report*, vol. 4, exhibit M-85, pp. 299–300.

I didn't fire a round yet and didn't kill anybody not even a chicken I couldn't. We are know suppose to push through 2 more it is about 10 A.M. and we are taken a rest before going in. We also got 2 weapons M1 and a carbine our final desti[na]tion is the Pinkville suppose to be cement bunkers we killed about 100 people after a while they said not to kill women and children. Stopped for chow about 1 P.M. we didn't do much after that. We are know setting up for the night 2 companies B and someone else we are set up in part of a village and rice patties had to dig foxhole area is pretty level are mortars are out with us. Are serving hot chow tonite I looked in my pack for dry socks and found out they were stolen from the time we were out in the field the name of the villages are My Lai 4, 5, and 6. I am know pulling my guard for night. 1 1/2 hours I am with the 1st squad had pop and beer. Sky is a little cloudy but it is warm out.

Mar 17 Sun: got up at 6:30 foggy out. We didn't go to Pinkville went to My Lai 2, 3, and 4 no one was there we burned as we pushed. We got 4 VC and a nurse. Had documents on them yesterday we took 14 VC. We pushed as far as the coast to the South China Sea there was a village along the coast also a lot of sailboats we stayed there for about an hour we went back about 2 kilometers to set up camp its in a graveyard actually we didn't pull guard but awake most of the night.

Mar 18 Mon: moved back to another area 1 VC said he would take us to a tunnel he took us all over didn't find any after that we met with other platoons as we were going 2 guys hit mines there flack jackets saved them not hurt bad Trevino and Gonzalez. . . . [T]here is a lot of fuss on what happened at the village a Gen was asking questions. There is going to be an investigation on Medina. We are not supposed to say anything. I didn't think it was right but we did it but we did it at least I can say I didn't kill anybody. I think I wanted to but in another way I didn't.

52

BRIAN LIVINGSTON

Letters Home

March 16 and 19, 1968

On the evening of March 16, 1968, Captain Brian Livingston, who commanded a group of assault helicopters, wrote a letter to his wife describing what he had seen at My Lai. Then, two days after the massacre, he read the description of the My Lai operation that appeared in the Americal News Sheet, *the newsletter of the Americal Division. The article outraged Livingston. The next day, he wrote again to his wife explaining some of his feelings.*

Saturday 16 March 68

Dear Betz,

Well its been a long day, saw some nasty sights. I saw the insertion of infantrymen and were they animals. The[y] preped the area first, then a lot of women and kids left the village. Then a gun team from the shark[s], a notorious killer of civilians, used their minny guns, people falling dead on the road. I've never seen so many people dead in one spot. Ninety-five percent were women and kids. We told the grunts on the ground of some injured kids. They helped them alright. A captain walked up to this little girl, he turned away took five steps, and fired a volly of shots into her. This Negro sergeant started shooting people in the head. Finally our OH23 saw some wounded kids, so we acted like medivacs. Another kid whom the grunts were going "take care of" was next on our list. The OH23 took him to Quang Nai hospital. We had to do this while we held machine guns on our own troops—American troops. I'll tell you something it sure makes one wonder why we are here. I can also see why they hate helicopter pilots. If I ever here a shark open his big mouth I'm going to shove my fist into his mouth.

We're trying to get the captain and sergeant afore mentioned reprimanded. I don't know if we will be successful, but we're trying. Enough for that.

Brian

Peers, *Report*, vol. 4, exhibit M-21, p. 111; exhibit M-22, p. 113.

19 March 68

Dear Betz,

You remember I told you about the massacre I witnessed, well I read a follow-up story in the paper. The article said I quote "The American troops were in heavy combat with an unknown number of V. C. Two Americans were killed, seven wounded, and 128 V. C. killed." Thats a bunch of bull. I saw four V. C., that is, those with weapons, and the amazing thing about that, is two of them got away. It made me sick to watch it.

Brian

53

SAMUEL KOSTER

Testimony to Peers Commission
1970

Major General Samuel Koster, commander of the Americal Division when the My Lai massacre occurred, had been named commandant of West Point by the time the Peers inquiry was under way. Koster denied that there had been a cover-up of My Lai, insisted that a thorough investigation of the incident had been conducted, and claimed that he had seen a formal, written report. Peers investigators were unable to locate such a document. Koster was formally interrogated in his West Point office by Lieutenant General William Peers.

PEERS: Well, we keep hearing, Gen. Koster, about a formal report. I say "keep hearing"; that is erroneous. I must rephrase that. I can say that except for Col. Henderson and yourself, to my knowledge at the moment we have been able to turn up absolutely nothing concerning any formal report that was ever made.

KOSTER: When I used the word "formal report," it was formal in the sense that these were written statements by a number of people who had

Michael Bilton and Kevin Sim, *Four Hours in My Lai* (New York: Penguin Books, 1992), p. 304.

been interrogated verbally prior to that time, and I know there was a sheaf of papers so big that it included these reports.

PEERS: We have talked to some of the people in command positions. They indicate that nobody ever talked to them. They made no signed statements. They gave no testimony under oath. We have talked to well over 70 people out of Charlie Company alone and we have yet to find one man who made a sworn statement or gave any testimony.

KOSTER: I'm not positive that the statements were sworn but they were signed.

PEERS: They made no statements and to further compound the problem there is no record of such a report ever having arrived at HQ, Americal Division. There is no copy of the report available. There is no information whatsoever aside from that which you and Col. Henderson have indicated. . . . We can find absolutely no one who has any knowledge of this report, including people who may have signed it, the people who may have prepared it, or any witnesses. We have talked to practically every individual who had any typing or preparation responsibility in the HQ of the 11th Brigade and we can find absolutely no reference to it.

KOSTER: I can't explain that.

PEERS: So if you did receive such a report, the way the evidence would appear at the moment, it would have been a complete forgery.

7

Exposure and Investigation

On March 29, 1969, Ronald Ridenhour, a Vietnam War veteran, wrote a letter to American political leaders outlining what he had heard about the events at My Lai and calling for a thorough investigation. The letter triggered formal inquiries in Congress and in the U.S. army. More than seven months later, on November 13, 1969, Seymour Hersh's investigative articles about My Lai appeared in newspapers around the country, as did a piece by Robert M. Smith in the *New York Times*. Early in December 1969, *Life* magazine published Ronald Haeberle's graphic color photographs of the massacre. The cover-up of the incident, which had been systematic at all levels of command in the Americal Division, was finally over. The entire world now knew about My Lai.

The revelations triggered a storm of controversy throughout the United States. For antiwar activists, exposure of the My Lai massacre and the army's cover-up provided incontrovertible proof that the American military effort in South Vietnam was immoral as well as misguided, that the United States was engaging in an orgy of violence and wreaking havoc on the civilian population. My Lai, they argued, was symptomatic of a culture of racism and violence that had become endemic in the United States.

Military spokesmen, as well as Nixon administration officials, claimed that My Lai was an isolated event, that although a catastrophe had been visited on one village, the incident was not representative of the entire war. American soldiers, they claimed, treated South Vietnamese civilians with dignity and worked among them to strengthen relations. After all, the argument went, the United States was in Indochina to protect the South Vietnamese from Communist atrocities.

But there was another defense as well. Many Americans wanted to exonerate Lieutenant William Calley and the troops of Charlie Company, not because the event had never occurred but because Vietnam was a different type of war. The soldiers had been given an impossible task—to win a war without being able to fully employ their resources; to confront an enemy that mixed inconspicuously with the civilian population; to fight a war without fronts in which they often found themselves shedding blood for the same territory over and over again; and to risk their lives when tens of millions of other Americans opposed the war. What happened at My Lai, according to these critics, was a tragedy, but the soldiers on the ground had not been responsible. If blame were to be attached to anyone, it should be firmly pinned on policymakers in Washington, D.C., who had led a naive nation down the road to Vietnam.

The entire controversy posed a real political problem for Nixon administration officials. They bore no responsibility for the massacre, of course, because My Lai had occurred long before Richard Nixon entered the White House. But significant political risks existed nonetheless. Antiwar liberals wanted to see the perpetrators of the massacre and the architects of the cover-up brought to justice. But antiwar liberals were not and had never been among Nixon's political supporters. Among southerners, blue-collar workers, and patriotic, pro-military Americans, support for the soldiers, if not for their military superiors, ran quite high. The political dilemma was obvious—how to satisfy the demands of justice without alienating some of the Nixon administration's strongest supporters.

54

RONALD RIDENHOUR

Letter to Military and Political Leaders

March 29, 1969

Ronald Ridenhour served with the aviation section of the 11th Infantry Brigade in South Vietnam from January through December 1968. Throughout the year he came into contact with other soldiers from the 1st Battalion of the 20th Infantry of the 11th Brigade. Many of them repeated to him rumors of the My Lai massacre; after his discharge from the army, Ridenhour felt compelled to get to the bottom of the rumors. On March 29, 1969, at his home in Phoenix, Arizona, he wrote a five-page, single-spaced letter to military and government leaders, including the secretary of defense and

Peers, *Report,* vol. 1, bk. 1, pp. 7–11.

*members of Congress. His letter was the first step in the public exposure of
what had happened at My Lai.*

Mr. Ron Ridenhour
1416 East Thomas Road #104
Phoenix, Arizona

March 29, 1969

Gentlemen:

It was late in April, 1968 that I first heard of "Pinkville" and what
allegedly happened there. I received that first report with some skepticism,
but in the following months I was to hear similar stories from such a wide
variety of people that it became impossible for me to disbelieve that
something rather dark and bloody did indeed occur sometime in March,
1968 in a village called "Pinkville" in the Republic of Viet Nam.

The circumstances that led to my having access to the reports I'm about
to relate need explanation. I was inducted in March, 1967 into the U. S.
Army. After receiving various training I was assigned to the 70th Infantry
Detachment (LRP), 11th Light Infantry Brigade at Schofield Barracks,
Hawaii, in early October, 1967. That unit, the 70th Infantry Detachment
(LRP), was disbanded a week before the 11th Brigade shipped out for Viet
Nam on the 5th of December, 1967. All of the men from whom I later
heard reports of the "Pinkville" incident were reassigned to "C" Company,
1st Battalion, 20th Infantry, 11th Light Infantry Brigade. I was reassigned
to the aviation section of Headquarters Headquarters Company 11th LIB.
After we had been in Viet Nam for 3 to 4 months many of the men from the
70th Inf. Det. (LRP) began to transfer into the same unit, "E" Company,
51st Infantry (LRP).

In late April, 1968 I was awaiting orders for a transfer from HHC, 11th
Brigade to Company "E," 51st Inf. (LRP), when I happened to run into Pfc
"Butch" Gruver, whom I had known in Hawaii. Gruver told me he had been
assigned to "C" Company 1st of the 20th until April 1st when he transferred
to the unit that I was headed for. During the course of our conversation he
told me the first of many reports I was to hear of "Pinkville."

"Charlie" Company 1/20 had been assigned to Task Force Barker in
late February, 1968 to help conduct "search and destroy" operations on
the Batangan Peninsula, Barker's area of operation. The task force was
operating out of L. F. Dottie, located five or six miles north of Quang Nhai
[*sic*] city on Viet Namese National Highway 1. Gruver said that Charlie
Company had sustained casualties; primarily from mines and booby traps,
almost everyday from the first day they arrived on the peninsula. One
village area was particularly troublesome and seemed to be infested with

booby traps and enemy soldiers. It was located about six miles northeast of Quang Nhai city at approximate coordinates B.S. 728795. It was a notorious area and the men of Task Force Barker had a special name for it: they called it "Pinkville." One morning in the latter part of March, Task Force Barker moved out from its firebase headed for "Pinkville." Its mission: destroy the trouble spot and all of its inhabitants.

When "Butch" told me this I didn't quite believe that what he was telling me was true, but he assured me that it was and went on to describe what had happened. The other two companies that made up the task force cordoned off the village so that "Charlie" Company could move through to destroy the structures and kill the inhabitants. Any villagers who ran from Charlie Company were stopped by the encircling companies. I asked "Butch" several times if all the people were killed. He said that he thought they were, men, women and children. He recalled seeing a small boy, about three or four years old, standing by the trail with a gunshot wound in one arm. The boy was clutching his wounded arm with his other hand, while blood trickled between his fingers. He was staring around himself in shock and disbelief at what he saw. "He just stood there with big eyes staring around like he didn't understand; he didn't believe what was happening. Then the captain's RTO (radio operator) put a burst of 16 (M-16 rifle) fire into him." It was so bad, Gruver said, that one of the men in his squad shot himself in the foot in order to be medivac-ed out of the area so that he would not have to participate in the slaughter. Although he had not seen it, Gruver had been told by people he considered trustworthy that one of the company's officers, 2nd Lieutenant Kally (this spelling may be incorrect) had rounded up several groups of villagers (each group consisting of a minimum of 20 persons of both sexes and all ages). According to the story, Kally then machine-gunned each group. Gruver estimated that the population of the village had been 300 to 400 people and that very few, if any, escaped.

After hearing this account I couldn't quite accept it. Somehow I just couldn't believe that not only had so many young American men participated in such an act of barbarism, but that their officers had ordered it. There were other men in the unit I was soon to be assigned to, "E" Company, 51st Infantry (LRP), who had been in Charlie Company at the time that Gruver alleged the incident at "Pinkville" had occurred. I became determined to ask them about "Pinkville" so that I might compare their accounts with Pfc Gruver's.

When I arrived at "Echo" Company, 51st Infantry (LRP) the first men I looked for were Pfc's Michael Terry and William Doherty. Both were veterans of "Charlie" Company, 1/20 and "Pinkville." Instead of contradicting "Butch" Gruver's story they corroborated it, adding some tasty tidbits of information of their own. Terry and Doherty had been in the same squad and their platoon was the third platoon of "C" Company to pass through the village. Most of the people they came to were already dead. Those that weren't were sought out and shot. The platoon left nothing alive, neither livestock nor people. Around noon the two soldiers' squad stopped to eat. "Billy and I started to get out our chow," Terry said, "but close to us was a bunch of Vietnamese in a heap, and some of them were moaning. Kally (2nd Lt. Kally) had been through before us and all of them had been shot, but many weren't dead. It was obvious that they weren't going to get any medical attention so Billy and I got up and went over to where they were. I guess we sort of finished them off." Terry went on to say that he and Doherty then returned to where their packs were and ate lunch. He estimated the size of the village to be 200 to 300 people. Doherty thought that the population of "Pinkville" had been 400 people.

If Terry, Doherty and Gruver could be believed, then not only had "Charlie" Company received orders to slaughter all the inhabitants of the village, but those orders had come from the commanding officer of Task Force Barker, or possibly even higher in the chain of command. Pfc Terry stated that when Captain Medina (Charlie Company's commanding officer Captain Ernest Medina) issued the order for the destruction of "Pinkville" he had been hesitant, as if it were something he didn't want to do but had to. Others I spoke to concurred with Terry on this.

It was June before I spoke to anyone who had something of significance to add to what I had already been told of the "Pinkville" incident. It was the end of June, 1968 when I ran into Sargent [sic] Larry La Croix at the USO in Chu Lai. La Croix had been in 2nd Lt. Kally's platoon on the day Task Force Barker swept through "Pinkville." What he told me verified the stories of the others, but he also had something new to add. He had been a witness to Kally's gunning down of at least three separate groups of villagers. "It was terrible. They were slaughtering the villagers like so many sheep." Kally's men were dragging people out of bunkers and hootches and putting them together in a group. The people in the group were men, women and children of all ages. As soon as he felt that the group was big enough, Kally ordered an M-60 (machine-gun) set up and the people killed. La Croix said that he bore witness to this procedure at least three times. The three groups were of different sizes, one of about twenty people, one of about thirty people, and one of about forty people. When the first group was put together Kally ordered Pfc Torres to man the machine-gun and open fire on the villagers that had been grouped together. This Torres did, but before everyone in the group was down he ceased fire and refused to fire again. After ordering Torres to recommence firing several times, Lieutenant Kally took over the M-60 and finished shooting the remaining villagers in that first group himself. Sargent La Croix told me that Kally didn't bother to order anyone to take the machine-gun when the other two groups of villagers were formed. He simply manned it himself and shot down all villagers in both groups.

This account of Sargent La Croix's confirmed the rumors that Gruver, Terry and Doherty had previously told me about Lieutenant Kally. It also convinced me that there was a very substantial amount of truth to the stories that all of these men had told. If I needed more convincing, I was to receive it.

It was in the middle of November, 1968 just a few weeks before I was to return to the United States for separation from the army that I talked to Pfc Michael Bernhardt. Bernhardt had served his entire year in Viet Nam in "Charlie" Company 1/20 and he too was about to go home. "Bernie" substantiated the tales told by the other men I had talked to in vivid, bloody detail and added this. "Bernie" had absolutely refused to take part in the massacre of the villagers of "Pinkville" that morning and he thought that it was rather strange that the officers of the company had not made an issue of it. But that evening "Medina (Captain Ernest Medina) came up to me ("Bernie") and told me not to do anything stupid like write my congressman" about what had happened that day. Bernhardt assured Captain Medina that he had no such thing in mind. He had nine months left in Viet Nam and felt that it was dangerous enough just fighting the acknowledged enemy.

Exactly what did, in fact, occur in the village of "Pinkville" in March, 1968 I do not know for <u>certain</u>, but I am convinced that it was something very black indeed. I remain irrevocably persuaded that if you and I do truly believe in the principles, of justice and the equality of every man, however humble, before the law, that form the very backbone that this country is founded on, then we must press forward a widespread and public investigation of this matter with all our combined efforts. I think that it was Winston Churchill [*sic*] who once said "A country without a conscience is a country without a soul, and a country without a soul is a country that cannot survive." I feel that I must take some positive action on this matter. I hope that you will launch an investigation immediately and keep me informed of your progress. If you cannot, then I don't know what other course of action to take.

I have considered sending this to newspapers, magazines, and broadcasting companies, but I somehow feel that investigation and action by the Congress of the United States is the appropriate procedure, and as a conscientious citizen I have no desire to further besmirch the image of the American serviceman in the eyes of the world. I feel that this action, while probably it would promote attention, would not bring about the constructive actions that the direct actions of the Congress of the United States would.

Sincerely,

/s/ Ron Ridenhour

55

WILLIAM WILSON

"I Had Prayed to God That This Thing Was Fiction"

1990

Ronald Ridenhour's letter set in motion a preliminary U.S. army investigation. Late in April 1969, William Wilson, an army colonel in the inspector general's office in Washington, D.C., was given the assignment to conduct a preliminary investigation into the alleged incident at My Lai. Wilson described the investigation in the following article more than two decades later.

In the early spring of 1969 I was an Army colonel recently assigned to the office of the inspector general in Washington, and I was not particularly happy about it; I have always disliked living in Washington, and I think that most infantry officers would rather serve with troops than investigate allegations about irregularities in procurement, which was most of what the IG's D.C. office did. Our job was to look into complaints sent to us from the Executive Branch or the Congress, and seven or eight fresh ones circulated in each morning's Read File. When the file came around one morning in March, it contained a lengthy letter from an ex-serviceman named Ron Ridenhour; he had sent copies to the President, twenty-three members of Congress, the Secretaries of State and Defense, the Secretary of the Army, and the chairman of the Joint Chiefs of Staff. Gen. William Westmoreland had forwarded a copy to our office with orders to investigate. . . .

After reading the letter four times, I picked up the phone for an appointment with Maj. Gen. William Enemark, the inspector general. His secretary said I could see him at ten o'clock. I read the letter again and waited. I remember insisting to myself that this was only hearsay; Ridenhour himself had noted that. I remember thinking that this could not be true. At ten I entered General Enemark's office, saluted, and explained that I had asked to see him regarding the letter from Ridenhour. The general observed that it painted a sordid picture of our forces and added that

"I Had Prayed to God That This Thing Was Fiction," *American Heritage*, February 1990, 44–53.

General Westmoreland and several congressmen were very upset. I requested assignment to the case, pointing out that I was the only investigator he had with infantry combat experience; I had served with the 101st Airborne in World War II. In my time I'd seen civilians killed, but those deaths had been accidental, and I thought it important to have someone who was confident that he could separate atrocities from wartime incidents requiring lethal force; our bombers, of course, killed women and children every day. But if the Pinkville incident was true, it was cold-blooded murder. I hoped to God it was false, but if it wasn't, I wanted the bastards exposed for what they'd done.

General Enemark rose from his chair, walked to the window, and stared at the busy Washington traffic passing in front of the Forrestal Building. After a while he said that he agreed that the investigation needed an officer with combat experience. He cautioned me that if he gave me the job, I would have to keep an open mind, and urged me to remember that everything in that letter was hearsay. He wanted me to move fast; if the information leaked before we got the facts, it would do a lot of damage to any subsequent disciplinary actions.

I assured the general that I would work as hard to disprove as to prove the allegations. He said that he'd give me his decision that afternoon, picked up his phone, and nodded that the discussion was over; I saluted and returned to my office.

The general approved my request in writing with a letter of instructions that included a requirement to keep the investigation and the disclosures confidential until I had completed the case and told him my conclusions. I conferred with my division chief, and he assigned a court reporter for the case. Smitty was near retirement, a cheerful D.C. local and pretty good company; that was very good luck because it turned out we were going to spend the next three months living cheek by jowl in motels. An investigator from the IG's office spends the bulk of his time interviewing witnesses, and our interrogations are sworn testimony taken down verbatim by the accompanying court reporter.

It helps if you can follow what the witnesses are talking about. I made a hasty study of available maps. . . . The information in the letter originated with four . . . men, and the first step was obviously to interrogate Ridenhour, then find the four men who'd provided the information.

In the best of all worlds Ridenhour would turn out to be crazy, but the tone of his letter—eloquent, thorough, and absolutely convincing—didn't leave much hope of that. What hope there was resided in the fact that the entire letter was hearsay. With luck the information would be garbled, exaggerated, or just plain wrong; that would be determined by my interviewing as many of the men who had actually been there as I could still find. The letter, dated March 29, 1969, indicated that Ridenhour was in Phoenix, Arizona. I needed details and addresses. Perhaps he knew the location of his four buddies. I kept repeating to myself: This is not a criminal investigation, there is no cross-examination, you are not trying to convict anyone, only to determine the facts.

There was no problem locating Ridenhour; I picked up the telephone and dialed Phoenix information. Ridenhour knew the location of the men who had told him of the atrocity: Charles Gruver was in Oklahoma City; Michael Terry was in Orem, Utah; Sp4 William Doherty was at Fort Hood, Texas; and Sgt. Michael Bernhardt was at Fort Dix, New Jersey. Army records indicated that Sgt. Lawrence LaCroix, another of Ridenhour's informants, was at Fort Carson, Colorado. I called for appointments and told Smitty to pack for a week and arrange open-ticket air transport starting with Phoenix.

An investigator from the Department of the Army inspector general's office does not have subpoena powers. While military witnesses can be ordered to appear before the IG investigator, civilians may or may not cooperate. I was relieved that all the men seemed eager to assist when I spoke with them on the phone; I told the witnesses that I'd contact them on my arrival. Meanwhile, I dissected Ridenhour's letter to extract the specific allegations and establish the routine questions to begin interrogations. I decided to inform the witnesses that the investigation had been ordered by General Westmoreland, because these men had served under him, and it was my impression that most soldiers had a great deal of confidence in Westmoreland. I also decided to wear full uniform and decorations to reveal immediately that they were talking to a combat soldier. Several of the men said that if I hadn't been wearing the Purple Heart— I'd been wounded at Normandy—and combat infantry badge, the information they gave me might have been different. In my experience those are two awards a combat infantryman recognizes and respects.

We were trained in the IG school on investigative procedures, administering oaths, and legal restraints on interrogations, but this seemed to be a particularly tricky case. I conferred with the division chief and Col. Clement Carney, the IG's judge advocate officer (military attorney), about pre-interrogation warnings. We decided not to use them; this was not a criminal investigation but one designed to determine the facts. Warning a witness at the beginning of the questioning that anything he said might be held against him would defeat the effort. I would not be placing the witness in jeopardy because if he did incriminate himself, the testimony could not be used since warnings were not given. This was to become extremely important. Each witness was told not to disclose or discuss his interrogation because of the damage it would do to any subsequent legal actions, although once I was on the road, I got the impression that a number of the men had put out the word that I was coming and had given their buddies a sense of what sort of person I seemed to be.

I set out for Arizona with mixed feelings about my chances. Ridenhour's letter was dated March 29, 1969, one year and thirteen days after the My Lai assault. General Westmoreland's office had acknowledged receipt to Ridenhour on April 12, stating an investigation was under way; Westmoreland had officially turned the case over to the inspector general on April 23, and I was about to interview Ridenhour in Phoenix on the twenty-ninth, one month to the day after the date of his letter. Considering mail time, distance, scheduling, and interoffice routing, the matter was being handled with dispatch; unfortunately, I was starting thirteen months late.

Ridenhour's interrogation was conducted at night, as were the majority of the interviews; most of the men worked or went to school by day. At forty-five, I was older than almost all of them. Ridenhour and I met in a downtown Phoenix hotel and went over his letter until I had assured myself that my questions to the witnesses would cover all the allegations. Ridenhour was an extremely impressive young man, and while his allegations were still only hearsay, he was depressingly convincing. The fact that he had stuck with it, pieced the thing together, and followed through remains extraordinary. Smitty and I left Phoenix the next morning for Orem, Utah, to interview Michael Terry on the first of May. I read to him his statement in Ridenhour's letter: ". . . Calley [2d Lt. William Calley] had been through before us and all of them had been shot but many weren't dead . . . it was obvious they weren't going to get any medical attention. . . . I guess we sort of finished them off."

Terry acknowledged that this was the information he had given Ridenhour. While this was an ugly incident and grievous error, I had no doubt Terry had tried to put these people out of their misery.

We left Orem for Fort Carson, Colorado, where I interviewed Sgt. Lawrence LaCroix on May 2. Ridenhour had talked to him in June 1968 at the USO in Chu Lai; he had claimed to have witnessed Calley's gunning down at least three separate groups of villagers. . . .

LaCroix said that [the] statement [in Ridenhour's letter attributed to him] was correct and also testified that at some point during the morning someone in a helicopter complained over the radio, "From up here it looks like a blood bath. What the hell are you doing down there?"

If there had been such a person, it was someone we had to track down; subsequent interrogations produced the name of Hugh C. Thompson, Jr., and allegations that he had threatened Calley's people with a machine gun to make them stop murdering the villagers. I very much wanted to meet this man, and I asked Washington to track him down.

Smitty and I caught an early flight for Oklahoma City and interrogated Charles Gruver on May 3; the contacts were moving as planned. I called the Washington office and told them to order Sgt. Michael Bernhardt, stationed at Fort Dix, to report to me on May 8 at the inspector general's office in Washington.

Gruver said he had indeed seen the three- or four-year-old wounded boy machine-gunned by the radio operator; the information he had given Ridenhour concerning Calley was hearsay, but he trusted the people who'd told him.

Our flight from Oklahoma City was late, so I did not interview Sp4 William Doherty at Fort Hood till the fifth of May. Doherty had been with Michael Terry when they entered My Lai 4 after the initial assault, and he confirmed that nearly every living thing in the village had been shot: cows, chickens, dogs, babies, and unarmed women. A lot of the people had been murdered in, or shot and subsequently dragged to, a drainage ditch. This made just too many witnesses for the tale of mass murders to have been conjured up out of whole cloth; a repugnant picture was forming in my mind, and I could see that Smitty was depressed too. Smitty was a quiet man, but his emotions were evident in his actions and expressions. When witnesses described the alleged murders, he lost his appetite and sometimes stared expressionlessly out into space. I normally tried to read the testimony of previous witnesses prior to new interviews, but with our schedule of packing, traveling, and interrogating as a daily repetitive routine, this was impossible. Transcribing testimony is not something done hurriedly, and Smitty had no time. We returned to Washington on the sixth, and Smitty's transcription of the mounting testimony was backing up. By this stage I was using maps to have the men plot what they could remember.

I interrogated Sergeant Bernhardt on the eighth. By that time I had the names of 60 to 70 percent of the men from C Company. Ridenhour's letter had described Sergeant Bernhardt's actions at My Lai 4: ". . . 'Bernie' had absolutely refused to take part in the massacre . . . he thought that it was rather strange that the officers of the company had not made an issue of it."

Bernhardt said this was true. He had entered the village after the action had started because Capt. Ernest Medina had sent him to the landing zone to check a booby trap. He summarized the day: "We met no resistance and I only saw three captured weapons. We had no casualties. It was just like any other Vietnamese village—old papa-sans, women and kids. As a matter of fact, I don't remember seeing one military-age male in the entire place, dead or alive. The only prisoner I saw was in his fifties."

Bernhardt testified that he saw Charlie Company doing strange things. The troops were going into hooches and shooting them up. They were gathering people in groups and shooting them. He was still astonished by his very presence in such a place that day: "It was point-blank murder and I was standing there watching it."

Bernhardt was an extremely cooperative witness, but he didn't always distinguish between what he had seen and what he had heard; under close questioning, a lot of his testimony amounted to hearsay.

There was a key witness in Georgia, Capt. Ernest Medina, the C Company commander, who was attending the Infantry Advanced Officers School, a nine-month career course. Medina was obviously upset by the questions and allegations. I asked him about the allegation that he had shot a woman lying on the ground. He admitted that he had shot a body that he assumed was dead, but when he turned to walk away, he thought he saw, from the corner of his eye, the woman's hand moving under her body. He fired because he thought she might have been preparing to throw a grenade. I believed Medina; he seemed to be a pretty sharp officer, he was planning a life in the service, he loved the Army, and he seemed very far from a monster.

What had this force been attempting to do? I notified the Washington office to summon Maj. Charles Calhoun from Fort Monroe, Virginia, for interrogation at the Forrestal Building on May 19. Major Calhoun had been the operations officer for Task Force Barker, and he summarized the plan of attack. . . .

Col. Oran K. Henderson had assumed command of the brigade immediately before the massacre; he was ordered to report to the inspector general's office on May 26 from U.S. Army headquarters in Hawaii. I asked him to sketch on my map the Song My operation as he remembered it. He denied being told anything about his troops wantonly killing large groups of civilians, and he denied a machine-gun confrontation between Thompson, the man in the helicopter, and Calley. He did remember Thompson telling him that his soldiers on the operation on the sixteenth were "like a bunch of wild men and were wildly shooting throughout My Lai." Henderson said, "I recall asking him if he knew what were the results of the infantry units he had supported." But first Henderson told Thompson that 20 civilians had been killed, and 128 Vietcong. Thompson replied that the bodies he saw on the ground were not VC but old men, old women, and children. Henderson had already received assurances from another officer about a "fierce fire fight," and he said that "when [Thompson] was talking to me he was in tears. . . . It appeared to me that the young warrant officer was . . . new and inexperienced."

Calley's platoon seemed to be the centerpiece of whatever had gone horribly wrong. I had interrogated SFC Isaiah Cowan at Fort Jackson, South Carolina, on May 23. Sergeant Cowan was Calley's platoon sergeant. The two had argued bitterly about tactics and procedures, Cowan said, but "Calley was my superior officer and I had to follow him whether I liked it or not . . . you have to go with your officers."

William L. Calley, Jr., had flunked out of Miami's Palm Beach Junior College in 1963. He'd moved west and worked for several years before enlisting in the Army in 1966. Despite a poor academic record he was selected for officer candidate school and graduated without learning to read a map properly. The consensus in the platoon boiled down to a single question: How had the Army ever considered Calley officer material? . . . I made arrangements for Calley to be returned to the United States from Vietnam and report to the inspector general's office on June 9.

Meanwhile, the Army found CW2 Hugh C. Thompson, Jr., and issued special orders for him to report to me on June 11 from Fort Rucker, Alabama. I kept him in Washington for three days while Thompson tried to accomplish, on the spot, what Ridenhour had done a year earlier: He attempted to reconstruct the entire incident. He was not entirely successful—and I do not think any honest eyewitness could have been—but he made prodigious efforts. He said that as he flew over the hamlet in his helicopter, he began seeing wounded and dead Vietnamese civilians everywhere. He decided to mark the location of wounded civilians with smoke so the GIs could start treating them. "The first one I marked was a girl that was wounded," Thompson testified, "and they came over to her, put their weapon on automatic and let her have it.". . .

Thompson had not been assigned to C Company; he was an outsider. I needed him to identify Calley—if Calley was indeed the man he had seen performing these acts—and I arranged for a lineup. I was nervous about prompting a witness, and I took some pains to explain to Thompson that all that we were trying to do was establish if one of these individuals was one of the people he'd talked to and that his identification might do as much to clear an individual as it would to accuse someone of having been present.

The next morning, June 13, Thompson picked out Calley from a lineup, identifying him as the officer at the drainage ditch in My Lai 4. Thompson also reported seeing a captain shoot a woman at close range while she lay on the ground. Without identifying Medina, I repeated his testimony about the movement of the woman's hand. Thompson said that "nothing is impossible." Thompson testified that he spent twenty to thirty minutes the following day telling Colonel Henderson his account of the massacre at My Lai 4. "I told him I had seen the Captain shoot the Vietnamese girl. I told him about the ditches and the bodies in the ditch. . . . I told him how I had gotten the people out of the bunker. I told him what I said to the Lieutenant."

Thompson estimated the number of bodies in the ditch at between seventy-five and one hundred. I found Thompson immensely impressive; he was the only hero of that awful day, and his testimony was damning. The trick would be to corroborate it.

This got easier when I found CW2 Dan R. Mullians at Fort Walters, Texas, and cut special orders for him to report to me in Washington on June 18. Mullians had been flying a helicopter in support of Charlie Company that morning, had heard no shooting and received no fire. He noticed the numerous bodies scattered in and around the village, in particular the ditch with bodies piled five to six feet deep. He stated that Thompson was enraged, and he had heard him say over the radio that "if he saw the ground troops kill one more woman or child he would start shooting [the ground troops] himself."

Then we had PFC Lawrence M. Colburn, Thompson's door gunner, brought to Washington from Fort Hood on June 19. He also confirmed Thompson's testimony. I showed him a number of photographs, and he picked out Medina and Calley as the officers involved in the shootings. He had gotten out of the aircraft and walked toward the ditch with the crew chief, who had then crawled into the ditch; the crew chief had been knee-deep in people and blood. Colburn vividly remembered that they had found one young child alive, buried under the bodies. "He was still holding on to his mother. But she was dead."

The boy, clinging desperately, was pried loose. Thompson said, "I don't think he was even wounded at all, just down there among all the other bodies, and he was terrified." Thompson and his men flew the baby to safety.

Maj. Glen D. Gibson, Thompson's company commander, was located at Headquarters 6th U.S. Army and ordered to report to Washington on June 25. There was a conflict between his and Colonel Henderson's testimony; Henderson had testified that sometime that evening he asked Major Gibson, commander of Thompson's aviation company, to look into this matter of his gunships threatening American soldiers and also whether the crews had observed "any of my soldiers shooting at civilians." Henderson stated that Gibson reported that "none of them had heard or seen any indiscriminate shooting, nor had they participated in any. He got a complete negative response from his people." Gibson persistently denied having any conversations with Henderson about My Lai 4.

The next step was to talk to more of Calley's people who had been on the ground. Ronald D. Grzesik was fire-team leader in Calley's platoon. On June 26 in Springfield, Massachusetts, I questioned him about his orders prior to the attack on My Lai 4. He had heard Medina tell the men "to go in and destroy the village; to make it uninhabitable," but he did not recall an order to destroy the inhabitants. For Grzesik, My Lai 4 was the end of a vicious circle that had begun months earlier: "It was like going from one step to another worse one.... First you'd stop the people, question them and let them go. Second, you'd stop the people, beat up an old man, and let them go. Third, you'd stop the people, beat up an old man and then shoot him. Fourth, you'd go in and wipe out a village."

While Grzesik was closing in on the village with his team, he was told to go to Calley. He walked over, and Calley ordered him to go to the ditch and "finish off the people." Grzesik refused, Calley asked him again, and Grzesik again refused. "I really believe he expected me to do it," Grzesik said with amazement. Calley angrily ordered him to take his men and burn the village. About three-quarters of the way through My Lai 4, Grzesik came upon an infantryman named Meadlo. It was after nine o'clock, and Meadlo was crouched, head in his hands, sobbing like a bewildered child. "I sat down and asked him what had happened." Meadlo said that Calley had made him shoot people. Grzesik felt responsible because he was the team leader.

I was struck by the picture of this man Meadlo, crying by the bodies of the dead; he was possibly the crucial witness, the last man I needed to present the truth of My Lai. Further interrogation disclosed that Meadlo had lost his foot to a land mine the following day. While being loaded on the evacuation helicopter, he shook his fist at Calley and screamed, "God will punish you for what you made me do!" He was evacuated and discharged from the service. That meant he would remember the day as no other man would, because his account of the activities would not be influenced by barracks-room discussions following the operation. I knew from my own experience that combat troops eventually come up with an agreed-upon version of violent and dramatic events, and men defer to the authority of an account they have unconsciously collaborated on. So Meadlo was the one.

I worked my way around the country. I interviewed Dennis R. Vazquez (formerly a captain) on July 1, in Williamsburg, Virginia. He was the task-force artillery liaison officer and testified that his forward observer reported sixty-nine Vietcong killed by artillery. To an experienced soldier, this had been a remarkably high body count for a three- to five-minute artillery barrage. I moved on to interrogate witnesses in Fort Worth, Texas, on July 9 and in Uvalde, Texas, shortly afterward and then in New Kensington, Pennsylvania, on July 15. I focused on a series of questions concerning Meadlo; I wanted confirmation of his remorse and his actions after the alleged atrocity. He became more important with each witness.

The evening of the sixteenth of July in a motel in Terre Haute, Indiana, is a time I would like to block from my memory. Paul D. Meadlo, his right foot and self-respect gone, came to the motel determined to relieve his conscience and describe the horrors of My Lai. He stated that Calley had left him and a few others with the responsibility of guarding a group of about eighty people who had been taken from their homes and herded together. He repeated Calley's instructions. "You know what I want to do with them," Calley said, and walked off. Ten minutes later he returned and asked, "Haven't you got rid of them yet? I want them dead. Waste them!" After telling about this, Meadlo raised his eyes to the ceiling of the motel room and began to cry. His compassion for the victims had taken control of him months before, and his body shook with sobs as he continued. "We stood about ten to fifteen feet away from them and then he started shooting. Then he told me to start shooting them. I used more than a whole clip—used four or five clips."

I was shocked. He blurted this confession out immediately after I asked him to tell me what happened at My Lai. I stopped him and told him to wait outside the room with Smitty. I called Washington, contacted Colonel Carney at home, and told him a witness had confessed to murder and I had not warned him of his rights. Colonel Carney instructed me to give Meadlo his warnings and see if he would repeat the confession. After he was warned that anything he said could be held against him in a court of law, he said, "I don't care." He repeated his confession.

I decided that the case was closed.

There was no doubt in my mind that a massacre had been committed at My Lai 4. Something in me had died as I watched Meadlo regress to the revulsion of the massacre at My Lai on March 16, 1968. I had prayed to God that this thing was fiction, and I knew now it was fact. I returned to Washington to report my findings on July 17, ten weeks after I had first interviewed Ridenhour. The report was to go to the chief of staff, the President, and Lt. Gen. William R. Peers. On August 19 I flew to Fort Benning to brief the legal officers there about the case. Under Army regulations, the commanding officer of Fort Benning and his legal staff were the ultimate authority for reviewing the evidence and filing charges against Calley.

On November 26, 1969, the Secretary of the Army and the chief of staff issued a joint memorandum directing General Peers to explore the nature and scope of the original Army investigations of what had occurred in Son My Village. The Peers inquiry was neither to include nor to interfere with the criminal investigation in progress.

Over my strong objections, I was ordered to participate in the inquiry; I stated, to no avail, that four months of this nightmare was enough. It was decided early that in order to discover the extent of the cover-up, the investigation must determine what had occurred in the entire Song My area on March 16 and 17, 1968. The Peers inquiry discovered an equally vicious massacre that had been conducted by a second company (B Company) of Task Force Barker on the same day. We reported that a part of the crimes visited on the inhabitants of Son My Village included individual and group acts of murder, rape, sodomy, maiming, assault on noncombatants, and the mistreatment and killing of detainees. Murders were common, as was discovered by the inspector general, the Peers inquiry, and the Criminal Investigation Division. Eventually some two dozen officers and men were charged by the Army with direct involvement in the killing at My Lai 4, but the only one convicted was Calley. In the words of General Peers, "The failure to bring justice to those who inflicted the atrocity casts grave doubts upon the efficacy of our justice system."

As for Calley, he was sentenced to life, which was reduced by the reviewing authority to twenty years and by the Secretary of the Army to ten. In the end he served three years under house arrest, and today he is working in his father-in-law's jewelry store in Columbus, Georgia.

I never made any particular effort to keep track of the witnesses from Charlie Company, but somehow you hear this or that, and sometimes it sticks. Meadlo is in Terre Haute, Indiana; he had trouble finding work with a foot blown off, but he eventually got a factory job making plastic film. I think he was laid off in November of 1988. Ridenhour became a writer; he now works for *City Business* in New Orleans, and in March 1988 he won the George Polk Award for local reporting after he'd ferreted out a lot of graft in the city government.

Mostly, though, I am struck by how little of these events I can or even wish to remember and how little any of us wanted to think about them at the time. I worked in Washington for many months on the matter of the massacres, surrounded by my fellow officers and friends. At no time, even long after the question of security had ceased to be of any conceivable relevance, do I remember ever talking to a single person about any of it. I do remember being startled when the public seemed to make a hero out of Rusty Calley, or at the least a victim. It sure didn't look that way from up close.

NEWSPAPER ACCOUNTS

Beginning on November 13, 1969, the My Lai massacre was front-page news in dozens of newspapers around the country. Seymour Hersh, a freelance Washington journalist, had sold his story about My Lai to the Dispatch News Service, which then distributed the story to thirty-five newspapers. While Hersh was writing his article, Robert M. Smith of the *New York Times* was writing a similar investigative report. The *Times* published his article about My Lai also on November 13. Coverage continued with more reports at the end of the month about the ongoing investigations.

56

ROBERT M. SMITH

Officer Kept in Army in Inquiry into Killing of Vietnam Civilians
November 13, 1969

WASHINGTON, Nov. 12—The Army has retained a 26-year-old first lieutenant on active duty for two and a half months beyond his term of service while it investigates charges that he shot "quite a number" of Vietnamese civilians.

Robert M. Smith, *New York Times,* November 13, 1969.

First Lieut. William Laws Calley Jr. of Miami, is being retained at Fort Benning, Ga. Officers at the post said that on Sept. 5 Lieutenant Calley was charged with the murder of an unspecified number of civilians in Vietnam in 1968.

Since then, the lieutenant has been kept on at the post as a special deputy to the deputy post commander, Col. T. W. Long.

The Army said that it would not disclose the specifics of the alleged crime that is being investigated "in order not to prejudice the continuing investigation or the rights of the accused."

Officials at Fort Benning are trying to determine whether there is sufficient evidence to convene a court-martial to try the lieutenant.

However, Lieutenant Calley's civilian attorney, George W. Latimer—who represented one of the soldiers involved in the recent Green Beret murder case—described the allegations against the officer.

He said that there were six specifications against the lieutenant and that they allege that he killed "quite a number of people." He would not specify how many.

Mr. Latimer explained that "some of the specifications might duplicate one another" and that, in any case, they are vague—"some of them say he killed 'not less than' so many."

Pressed about the number of people allegedly killed, Mr. Latimer said, "I would guess that if all the specifications were added together the number might reach 109." One hundred nine had been suggested to him as the figure being commonly cited in reports from Fort Benning.

The specifications—as Mr. Latimer related them—paint this picture of the alleged crime:

Lieutenant Calley was in command of a platoon of Americal division soldiers of the 11th Infantry Brigade. The platoon was part of a task force ordered in March, 1968, to advance on and take several villages in a Vietcong stronghold called Pinkville.

The lieutenant's platoon was to take one village, or perhaps a few villages strung together, six miles northeast of Quangngai. The area had been under heavy shelling by Navy gunboats and ground artillery—"a lot of firepower," Mr. Latimer added.

It is alleged that the lieutenant advanced with his troops into the village and, with premeditation, killed what Mr. Latimer called "a multiple number" of civilians with his rifle.

Mr. Latimer said that the lieutenant was not guilty, "it was in the line of duty," he said.

The lawyer—who was a judge on the United States Court of Military

Appeals for 10 years and now has a law practice in Salt Lake City, Utah, said the case reminded him of the recent Green Beret affair, in which American servicemen were accused—but finally free—of killing an alleged Vietcong double agent.

"We are trying our own people for shooting people who were in the enemy camp," Mr. Latimer said.

"They were Vietcong villagers," he added, and these people worked for the Vietcong—carrying ammunition, and so on."

The lawyer also said that "half a dozen others may be involved." When asked to be more specific, he demurred—acknowledging only that he was speaking of the lieutenant's superiors.

The Army said that Lieutenant Calley was under no restraint at Fort Benning and was "on a full-duty status."

Pentagon officials said early last night that the investigation had reached the stage where the lieutenant's commander at Fort Benning now had to decide whether he should order the convening of a court-martial.

The Pentagon spokesman also said that all of the specifications made during the investigation may or may not be brought before a court-martial, if one is convened.

The Army spokesman likened the investigative proceedings to civilian grand jury hearings, with the decision to convene a court-martial parallel to the issuing of an indictment. In the military, however, the individual is charged before the investigation is complete.

If the investigation does not turn up enough evidence to warrant a court-martial—in the view of the commander—the charge is dropped.

57

E. W. KENWORTHY

Resor Called to Testify
about Alleged Massacre
November 26, 1969

WASHINGTON, Nov. 25—The Armed Services Committees of the Senate and the House of Representatives today summoned Secretary of the Army Stanley R. Resor to testify tomorrow on the alleged mass killing of South Vietnam villagers in Quangngai Province in March, 1968.

Mr. Resor, who was also Secretary of the Army when the alleged shooting took place, will be accompanied by Robert E. Jordan 3d, general counsel of the Army.

Meanwhile, in the first direct comment by a high Administration official, Secretary of Defense Melvin R. Laird said that he had been "shocked and sick" when he first heard of the alleged shooting 11 months after it was said to have occurred.

In New York, Representative Gerald R. Ford, the House Republican leader, said tonight that the attack at Songmy "was known about by top Army officers." However, he said, "I don't have it first hand" or, he added, "know them by name."

[Late Tuesday, Varnado Simpson, now a civilian, said on National Broadcasting Company television that he had participated in an attack on Songmy and was "personally responsible" for the deaths of 10 Vietnamese.]

In a telephone interview today, Clark M. Clifford, who was Secretary of Defense at the time of the alleged massacre, said:

"I had never heard of it before until the story broke in the newspapers. It had never been brought to my attention. I am assuming that it had never been brought to the attention of the Secretary of the Army, because if it had been, I think he would have taken it up with me."

Mr. Clifford went on to say: "As far as the facts are concerned, I will wait until they are subjected to the two important test[s] of evidence taken under oath and a searching cross-examination."

Both of Army Secretary Resor's appearances tomorrow [are] scheduled for mid-morning. Immediately after the Senate panel hears from the Secretary, he is to appear before the House Armed Services Committee's

---

E. W. Kenworthy, *New York Times,* November 26, 1969.

investigating subcommittee, along with a number of other Army witnesses, including Lieut. Gen. Richard Stilwell, Deputy Chief of Staff for military operations.

At Fort Benning, Ga., the Army has announced that First Lieut. William L. Calley Jr. would be given a general court martial on charges of premeditated murder. In the specifications, Lieutenant Calley is charged with killing—"unlawfully" and "without justification or excuse"—109 men, women and children in the village of Songmy on or about March 16, 1968.

Secretary Laird made his comment about being "shocked and sick" about the alleged massacre in reply to a question by Senator J. W. Fulbright during a seven-hour session with the Foreign Relations Committee last Wednesday. Secretary Laird did not answer the question immediately but sent a written reply for insertion in the hearing record. The reply was made public by Senator Fulbright today with Mr. Laird's consent.

In the Senate today Senators Peter H. Dominick, Republican of Colorado, and Ernest F. Hollings, Democrat of South Carolina, strongly criticized the Columbia Broadcasting System for carrying an interview with a former member of the infantry unit that allegedly shot the unarmed civilians during a sweep of Songmy.

The veteran interviewed was Paul Meadlo, 22 years old, who had lost a foot the day after the alleged massacre at Songmy. Mr. Meadlo said in the interview that, at the direction of Lieutenant Calley, "about 370" villagers had been killed; that he himself had fired about 67 shots and "might have killed 10 or 15." At the end of the interview, Mr. Meadlo said, "I see women and children in my sleep. Some days . . . some nights, I can't even sleep. I just lay there thinking about it."

Commenting on the broadcast in a floor speech, Senator Dominick contended that the legal rights of Lieutenant Calley and Mr. Meadlo had been jeopardized. The Senator said the broadcast interview was "in total disregard of the rules of the Supreme Court" regarding pretrial release of information in criminal cases.

"He [Meadlo] said he had personally participated in the murder of some of these men, some of these women and some of these children," Senator Dominick said. "He specifically mentioned the name of the man who is under indictment [Lieutenant Calley]. What kind of country have we got when this kind of garbage is put around?"

Senator Hollings rose in the nearly empty chamber and asked whether every soldier who had committed "a mistake in judgment" during the heat of combat was "going to be tried as common criminals, as murderers?"

Senator Hollings said that Mr. Meadlo "was obviously sick," and that a man in his condition "ought not to be exposed to the entire public."

Secretary Laird told the committee that, under the American system of jurisprudence and the provisions of the Uniform Code of Military Justice, he could not comment on the alleged shootings or on Lieutenant Calley and Staff Sgt. David Mitchell of St. Francisville, La., who has been charged with assault with intent to kill 30 persons. No decision has yet been announced as to whether Sergeant Mitchell will stand trial.

However, Mr. Laird said he wanted "to make clear, beyond any doubt, that the Nixon Administration is determined to insure absolute compliance with our orders and with the laws of war."

Secretary Laird told the committee that he had not heard about the alleged murders until early April of this year. This was a month after Ronald Ridenhour, a 23-year-old student who had been a soldier in Vietnam, wrote to the President, Mr. Laird, 23 members of Congress and a half dozen other officials. In his letter, he set forth what a friend had told him of the alleged Songmy massacre by members of the First Platoon, C Company, 1st Battalion, 20th Infantry of the 11th Infantry Brigade.

Mr. Laird gave no reason to the committee why the alleged incident had not been brought to his attention sooner by the Army, nor why the report of an investigation made by the 11th Brigade in April, 1968, had not been sent to the Department of the Army in Washington.

This report, according to the Army, concluded that no disciplinary action was appropriate and no further action was required. In a statement provided to the Senate Foreign Relations Committee today, the Army said: "The matter [meaning the alleged incident and the original investigatory report] was not brought to the attention of the Department of the Army [in Washington], there being no apparent requirement for doing so."

The Army has refused to say who conducted the original investigation in April 1968. But yesterday it announced that Secretary Resor had named Lieut. Gen. William R. Peers "to determine the adequacy of both the investigation and its subsequent review.". . .

Says He Took Part

In an interview broadcast on National Broadcasting Company television last night, Varnado Simpson, who is now a civilian, said he had participated as an enlisted man in the attack on Songmy.

Mr. Simpson of Jackson, Mich., said he was "personally responsible" for the death of 10 Vietnamese in the village, including an old man, a woman and a two-year-old child.

Asked why he had shot the civilians, Mr. Simpson replied that he was following orders. He said he had emptied 16 rifle magazines containing 20 bullets each.

Evidence Banned

COLUMBUS, Ga., Nov. 25 (AP) — A military judge ordered today that potential witnesses in the court-martial of First Lieut. William L. Calley Jr. be directed not to disclose any evidence they may have prior to the trial.

Lieutenant Calley has been charged by the Army with premeditated murder in the deaths of at least 109 Vietnamese civilians.

The directive concerning potential witnesses was issued by Lieut. Col. Reid Kennedy, senior trial judge in the judge advocate section at Fort Benning.

Colonel Kennedy's directive came at a hearing here. He had been asked by both the defense and prosecution attorneys to ban further news interviews with persons who might appear at the court-martial.

C.B.S. Reply

Following is a reply of the Columbia Broadcasting System to the attack on it by Senator Dominick for having broadcast the interview with Mr. Meadlo:

"C.B.S. News broadcast the interview with Paul Meadlo in the belief there was an overriding public need for full disclosure about what happened at Mylai, particularly in view of previous statements made by other eyewitnesses and then the statement issued by the Government of South Vietnam that nothing untoward had happened there.

"This South Vietnamese official position was then contradicted by the United States Army decision yesterday to try an American officer on charges of premeditated murder at Mylai.

"C.B.S. News believes that Paul Meadlo was entitled to make his story public if that was his decision, and having established to our satisfaction that Paul Meadlo was qualified to speak on the subject as a bona fide participant in that incident, we would be guilty of not reporting information to which the American public was entitled.

58

FRED P. GRAHAM

Army Lawyers Seek Way
To Bring Ex-G.I.'s to Trial

November 26, 1969

WASHINGTON, Nov. 25—Pentagon lawyers are searching lawbooks to find if there is any way to prosecute men who took part in what has been called the massacre at Songmy but who have since been released from active duty.

A military source disclosed today that if the Army's legal experts conclude that they can press charges against the discharged men—despite a 1955 Supreme Court ruling that they concede makes such prosecutions highly unlikely—an announcement will be made in the next few days.

The reason is that some participants in the incident who are no longer in uniform are giving interviews in the belief that they are immune from prosecution. This could be highly self-incriminatory if an effort is to be made to try them in connection with the reported killing of a large number of Vietnamese civilians in the village of Songmy in March of 1968.

If the Army concludes that it cannot try the men, this could prompt Congress to pass a law to permit the trial of former servicemen in civilian courts for crimes they committed while in uniform—a step that the Supreme Court suggested by implication in its 1955 ruling.

The ruling concerned Robert W. Toth, a Pittsburgh steelworker who was seized in his home in a midnight arrest by military policemen five months after his discharge from the Air Force. They took him to Korea to face charges of having murdered a Korean.

The Supreme Court held that any former serviceman who had completely severed his ties with the service could not be denied his constitutional right to a civilian trial. But Justice Hugo L. Black stated in the majority opinion that Congress did have the constitutional authority to establish civilian courts to try former servicemen under such circumstances.

Although Congress has amended the Uniform Code of Military Justice since then, it did not take this step. Pentagon spokesmen have said that of the 24 men now being investigated in connection with the incident, 15 have been released from active duty.

Fred P. Graham, *New York Times,* November 26, 1969.

Arthur J. Keefe, an expert on military jurisdiction, who teaches law at Catholic University Law School here, said in an interview here today that some of these men might still be subject to court-martial, if they still have ties to the military.

Cites Such a Case

In one case, a former soldier who was still in the inactive reserves was called back to active duty five months after his release and was court-martialed for a murder committed while in uniform.

Professor Keefe said that at least one retired officer had been court-martialed on the strength of his military pension. If any of the men have been released and have reenlisted, they might also be tried before military courts, he said.

Professor Keefe said he would consider this a bad policy—even if it were good law—and the Pentagon officials say privately that they are moving on the same assumption. All men who have left active duty will be treated alike, they say. Unless a way can be found to try all who appear culpable, none will be tried.

Some lawyers believe that Congress could pass a law even now to permit the trial of the ex-soldiers in Federal district courts in the United States. The Constitution forbids ex post facto laws that make certain acts crimes after they have been committed. But a law that created a court to try crimes that existed in military law at the time of the massacre might not be held to be an ex post facto law.

Military sources have made it clear that they are not absolving any participants on the theory that they merely took orders.

Soldiers are required to obey all "lawful" orders, but not obviously unlawful ones. There have been instances in which men have been court-martialed for crimes they committed under orders from superiors, Pentagon sources say.

59

Photographs from Life Magazine 1969

If it wasn't already, the My Lai massacre became embedded in worldwide public consciousness with the December 5, 1969, publication of Ronald Haeberle's color photographs in Life *magazine. The publication in* Life *precipitated a media frenzy as reports about the massacre appeared in newspapers and on radio and television news shows around the world.*

Life, December 5, 1969, 36–45.

PUBLIC REACTION

Public reaction to the massacre was predictable. Antiwar liberals condemned My Lai as symptomatic of everything that was wrong with the Vietnam War. Conservatives were much more cautious about what had happened at My Lai, and they were quite critical of liberals who rushed to judgment. The following two documents illustrate those reactions. The perspective of antiwar liberals is crystal clear in the "Notes and Comment" column from *The New Yorker*. A more conservative tone is set in "The Great Atrocity Hunt," in the *National Review*.

60

Notes and Comment
December 20, 1969

The reports of the massacre in the South Vietnamese village of My Lai, in which American troops are said to have rounded up several hundred villagers and then gunned them down, have left the nation stunned and vexed. We sense — all of us — that our best instincts are deserting us, and we are oppressed by a dim feeling that beneath our words and phrases, almost beneath our consciousness, we are quietly choking on the blood of innocents. Often, when we open our mouths to condemn, excuses pour out uncontrollably instead. When soldiers of other nations did such things, our outrage was clear and strong, and we had no trouble finding the words of condemnation, but now we find that that outrage was poor preparation for facing what appears to be an atrocity our own people have committed. We try to turn the old phrases of condemnation against ourselves, but they seem to ring false — perhaps because they rested in the first place on a complete dissociation of ourselves from the people we condemned — and the beginnings of a new, craven logic that finds such atrocities to be the way of the world steal into our minds. When others committed them, we looked on the atrocities through the eyes of the victims. Now we find ourselves, almost against our will, looking through the eyes of the perpetrators, and the landscape seems next to unrecognizable. The victims are indistinct — almost invisible. A death close to us person-

"Notes and Comment," *New Yorker,* December 20, 1969, 27.

ally seems unfathomably large, but their deaths dwindle in our eyes to mere abstractions. We don't know what kind of lives they led or what kind of things they said to each other. We are even uncertain of the right name of the village we are said to have annihilated. Our attention turns to the men who are charged with the crime. Could *we ourselves* have committed it? (Already we have dissociated these men from "us.") Explanations of how such things can happen which never occurred to us when others did them are suddenly ready to hand, and we try to use them to comfort ourselves. From the President on down, we have responded in a muted, tired way. Some people look to the trials of Lieutenant William Calley, Jr., and Sergeant David Mitchell to resolve the issue. But the massacre — if, indeed, there was a massacre — has raised questions that go far beyond the question of the guilt of the men charged with being participants, and although some of these broader questions may be raised at the trials, the only proper purpose of the trials is to decide the fate of the accused individuals. If the men accused are convicted, the question of how much responsibility the rest of us bear will be left to us, and will be resolved by what we say to each other and what we say to ourselves, and, above all, by what course we follow thereafter in our Vietnam policy.

Although it may be that the My Lai massacre is an "isolated incident," in the sense that no other report of mass killing of civilians by troops on the ground has been brought to light, there can be no doubt that such an atrocity was possible only because a number of other methods of killing civilians and destroying their villages had come to be the rule, and not the exception, in our conduct of the war. And the scale of this killing and destruction had been great enough even at the time of the My Lai massacre to defeat completely our original purpose in going into Vietnam, which was to *save* the South Vietnamese people from coercion by the enemy. A report in these pages revealed that in August, 1967, in Quang Ngai Province — where the alleged My Lai massacre took place seven months later — a carefully devised and clearly articulated policy of reprisal bombings against villages that helped the Vietcong, or even tolerated their presence under duress, was widely in effect. This reprisal policy was announced to the South Vietnamese population in the clearest possible terms by the dropping of psychological-warfare leaflets. One such leaflet, the title of which was "Marine Ultimatum to Vietnamese People," announced, in part, "The U.S. Marines issue this warning: THE U.S. MARINES WILL NOT HESITATE TO DESTROY, IMMEDIATELY, ANY VILLAGE OR HAMLET HARBORING THE VIETCONG. WE WILL NOT HESITATE TO DESTROY, IMME-DIATELY, ANY VILLAGE OR HAMLET USED AS A VIETCONG STRONGHOLD TO FIRE AT OUR TROOPS OR AIRCRAFT." The same leaflet announced the names of hamlets that had already been bombed, saying, "The hamlets of Hai Mon, Hai Tan, Sa Binh, Tan Binh, and many others have been destroyed." And, in case the villagers didn't fully understand us after they had read this—or, for that matter, after their village had been bombed—another leaflet, titled "Your Village Has Been Bombed," was dropped, telling why we had bombed it. At one point, the second leaflet informed the surviving villagers, "Your village will be bombed again if you harbor the Vietcong in any way." This policy did not result in a few isolated incidents of villages' being bombed; it resulted in the razing to the ground of over seventy per cent of the villages in the province. . . .

61

The Great Atrocity Hunt
December 16, 1969

Murder may have been committed at Songmy; nevertheless, what actually happened remains unclear. Statements made before TV cameras and interviews granted to itinerant reporters do not establish facts, and we will not know what happened before the evidence is assembled and the testimony, given under oath, is subjected to cross-examination. Every American serviceman arriving in Vietnam is instructed that "war crimes" include the "killing of spies, or other persons who have committed hostile acts, without trial," and Article 118 of the Uniform Code of Military Justice does not distinguish in the definition of murder between killing an American and killing a non-American civilian or prisoner. On the other hand, as everyone knows, a guerrilla war such as is being waged in Vietnam is full of ambiguities: if the Vietcong chooses to build its stronghold under a village, and the villagers, either through choice or coercion, continue to reside in the village, some perhaps to cooperate with the VC, does that mean the stronghold is immune from attack? The judicial process will establish whether a crime has been committed; if the answer is yes, the guilty, if still living, will be punished. Why is there any need, just now, to say more than that?

Whether atrocities were committed at Songmy we do not as yet know;

"The Great Atrocity Hunt," *National Review,* December 16, 1969, 1252.

but more than enough atrocities against human reason have been committed in response by the American media. That "America and Americans must stand in the larger dock of guilt and human conscience for what happened at Mylai seems inescapable." So observed *Time,* adding: "Men in American uniforms slaughtered the civilians of Mylai, and in so doing humiliated the U.S. and called in question the U.S. mission in Vietnam in a way that all the antiwar protesters could never have done." One's mind staggers: Are "America and Americans" generally guilty, the same America and Americans now preparing to bring the accused to trial? And even if it turns out that atrocities were committed, how does this call in question the U.S. mission in Vietnam? Do the innumerable atrocities committed by both sides in World War II add up to the proposition that resistance to the Nazis ought to have been abandoned? On April 9–10, 1948, Israeli commandos massacred all 254 inhabitants of the Arab village of Deir Yassin—men, women, and children. Does *Time* conclude from this that the Israeli position, in toto, is morally indefensible? During the American Civil War atrocity was not an aberration, the act of bewildered or temporarily unbalanced men, but a matter of settled military policy. "Until we can repopulate Georgia," said General Sherman, "it is useless for us to occupy it; but the utter destruction of its roads, houses, and people will cripple their military resources." Does *Time* conclude that the Union, therefore, should have been permitted to disintegrate? Irrational and irresponsible comment on Songmy has become collective madness.

Ronald Lee Ridenhour, the ex-GI, now a college student, who brought Songmy to the attention of thirty Washington officials, described it as a "dark" event—as it may well have been; but, most assuredly, there is something dark and sick about much of the reaction from the liberal Left. *N.Y. Post* columnist Pete Hamill, for example, quoted President Nixon's education speech to the effect that his Administration would be committed "to the enriching of a child's first five years of life," and then commented: "The long weekend is over now, in the nation that enriches children by machine-gunning them to death. . . . Let's get on to the next hamlet, boys: we've got some four-year-old Vietcong sympathizers to enrich. We're going to give them The American Way at seventeen rounds a clip." Mary McGrory's syndicated Washington column announced: "The country's conscience, apparently, died in that Asian village with the old men, the women and the children." (How did *her* conscience manage to survive the sudden mortality of American conscience?)

8

Culpability

Fixing blame, determining guilt, and punishing the perpetrators of the My Lai massacre proved to be far more complex than anyone first anticipated. Nearly five hundred Vietnamese civilians had been slaughtered on March 16, 1968, in front of dozens of eyewitnesses, but when all of the investigations were over, the indictments rendered, and the trials held, only Lieutenant William Calley was found guilty of war crimes. Calley was convicted of the premeditated murder of twenty-two civilians and sentenced to life in prison at hard labor. Colonel Oran K. Henderson, commander of the 11th Infantry Brigade, and thirteen other officers and enlisted men had been charged with war crimes, but each of them was acquitted or had his charges dropped. The only other punishments meted out were to military brass for the cover-up. Major General Samuel Koster was reduced in rank to brigadier general, and his assistant, Brigadier General George Young, had an official censure placed in his personnel file.

To antiwar liberals, the failure to convict any others for war crimes seemed a gross miscarriage of justice. A mountain of evidence, little of it circumstantial, seemed to prove guilt beyond reasonable doubt, but the six-man military tribunals did not see it that way. In each of the trials, most of the men sitting in judgment had themselves completed at least one tour of duty in Vietnam. In Captain Ernest Medina's trial, they deliberated for only an hour before acquitting him. A similar process took place in the other cases that finally made it to court. When no convictions were secured, charges against others were dropped. Why were the men serving on the tribunals unwilling to convict? Was it because they had seen so many atrocities in Vietnam that to convict the soldiers of Charlie Company would be an act of hypocrisy? Were they thinking "there but for the grace of God go I"? Or were they trying to protect the army's reputation by denying most of the charges? William Calley was convicted because the evidence against him was so egregious and overwhelming, but even then the guilty verdict ignited a storm of controversy. Many Americans were convinced that Calley had been set up as a scapegoat, that he was taking the fall for higher-ranking officers who had also participated in the atrocity. Governors George Wallace of Alabama and Jimmy Carter of Georgia, for example, claimed that Calley was only the "tip of the iceberg." Others held Calley blameless, arguing that the confusing nature of combat in Vietnam, where civilians and enemy troops were indistinguishable, made it impossible to observe the niceties of international law. Finally, some Americans saw Calley as a victim of an oppressive federal government; in doing so, they vested him with heroic status, or perhaps even made him another of the era's antiheroes.

Calley's conviction became a political football. The American Legion and the Veterans of Foreign Wars helped cover Calley's legal expenses. They lobbied members of Congress, senators, and the White House demanding parole, pardon, or clemency for Calley. Many Americans— especially in the South, Midwest, and West—viewed Calley as just another victim of My Lai, not a war criminal. The Nixon administration soon sided with Calley, gradually reducing his life sentence until November 9, 1974, when Secretary of the Army Howard Calloway paroled him. Upon hearing the news, Lieutenant General William Peers was dumbfounded. "To think that out of all those men," he told reporters, "only one, Lieutenant William Calley, was brought to justice. And now, he's practically a hero. It's a tragedy."

62

J. ANTHONY LUKAS

Meadlo's Home Town Regards Him
as Blameless
November 26, 1969

The people of New Goshen, Indiana, refused to blame Paul Meadlo, one of their own, for what happened at My Lai. The following article describes the community's reaction to news that Meadlo had participated in the massacre.

J. Anthony Lukas, "Meadlo's Home Town Regards Him as Blameless," *New York Times,* November 26, 1969.

NEW GOSHEN, Ind., Nov. 25—This village with the Biblical chime to its name woke up this morning as the home of the man who said he took part in the mass slaying of Vietnamese civilians.

But nowhere among its chunky church spires and white clapboard houses was anyone inclined to blame Paul David Meadlo, the veteran who admitted last night shooting 30 to 40 men, women and children in the massacre at Song My, South Vietnam, last year.

"Lots of people been in talking about it this morning," said Mrs. Josephine Neview, a clerk at Neal's grocery. "But they certainly don't blame Paul David in any way. After all, he had his orders."

"Paul David" is the way everybody here refers to the 22-year-old coal miner's son.

"I heard them announcing something about Paul Meadlo on the TV last night," said Mrs. Neview. "For a moment I didn't know who they were talking about; then I said, oh, my God, that's Paul David."

Although Paul David lives now with his wife and two children in neighboring West Terre Haute, New Goshen was his home for the first 18 years of his life, and he still comes here often to visit his parents, sister, four uncles, and two aunts.

New Goshen's 450 residents all know the Meadlos and Paul David; when he's home, he is a familiar figure shouldering up to the bar at Hutch's Hut or warming his hands over the pot-bellied stove at Olivero's Grocery.

And he is popular here. Townspeople questioned today responded with one voice: "A very nice boy," "the nicest guy you'd ever want to meet," "easy-going, got along with everybody," "never had any trouble out of him. Wish I could say the same about some other youngsters around here."

So when newspaper and television people from New York, Chicago, and St. Louis began showing up around town this morning, the people of New Goshen stuck staunchly by their native son.

"How can you newspaper people blame Paul David?" asked Robert Hale as he planed down some garage doors behind the pool hall he and his wife run. "He was under orders. He had to do what his officer told him."

"The only thing I blame Paul David for was talking about this to everybody on television," said Dee Henry, who was helping Mr. Hale fix the garage doors. "Things like that happen in war. They always have and they always will. But only just recently have people started telling the press about it."

"It's bad enough to have to kill people without telling everybody about it," he said. "This sort of thing should be kept classified."

Mr. Henry was a professional soldier for 11 years, fought in World War II and the Korean war, and "would have been in this one too if I hadn't been wounded and discharged." He gives his occupation as "disabled veteran."

He feels Paul David is the victim of people who don't know how the Army works. "Anybody who's had any affiliation with the service knows you do what you're ordered to do—no questions asked."

Although others were not so emphatic, the same theme was echoed today in the town's tiny one-room post office, on front porches, and on street corners under gaunt, leafless maples.

"What else could he do?" asked 22-year-old Floyd Cheesman, a classmate of Paul David's at West Vigo High School. "Boy, I would have done just the same thing. I'd take my orders. If they give you an order under fire and you don't carry it out, they can court martial you."

The only sharply differing point of view expressed in town today came from Paul David's father, a blunt ex-miner who still speaks with a trace of the Polish accent his father brought from the old country 67 years ago.

"If it had been me out there," he said, "I would have swung my rifle around and shot Calley instead—right between the God-damned eyes." Lieut. William L. Calley Jr. is the officer Paul David says ordered him to shoot the Vietnamese.

"Why did they have to take my son and do that to him?" said Mr. Meadlo's wife, Myrtle, as the tears she'd held back all morning began to flow.

Paul David came home without his right foot—blown off by a land mine the day after the Songmy massacre. His father stomps around the house on an artificial left leg—the result of a mine accident in 1961.

When Paul David first spoke to his mother from an Army hospital after the injury, he said, "Well, Mom, like father like son."

The Meadlos don't talk much about the victims of their son's gunfire in Songmy. But at times they seem to be thinking of them.

Showing a visitor a picture of Paul, his wife, Mary, and his two babies, Paul Jr., 2½, and Tresa Lynn, 15 months, Mrs. Meadlo said through her tears: "When he's around his babies he'll pick 'em up and love 'em. Just love 'em."

63

WILLIAM L. CALLEY

Testimony at Court-Martial

1970

At his court-martial trial on charges of committing war crimes, Lieutenant William Calley never denied that he had participated in the killing of non-combatants. Instead, he based his defense on the fact that he had been given orders to commit the killings and that he had obeyed those orders. In the following section of testimony, Calley responds to questions posed by his defense attorney.

Q: Now I will ask you if, during those three periods of instruction and training, were you instructed by anybody in connection with the Geneva Conference?

A: Yes, sir, I was.

Q: What was—if you have a recollection—what was the extent and nature of that tutoring or training?

A: I know there was classes. I can't remember any of the classes. Nothing stands out in my mind as to what was covered in the classes, sir.

Q: Did you learn anything in those classes of what actually the Geneva Conference covers with respect to the rules of warfare?

A: Not in the laws and rules of warfare, sir.

Q: Did you receive any training in any of those places which had to do with the obedience to orders?

A: Yes, sir.

Q: And what was the nature of the training, and what were you informed was the principles involved in that field?

A: That all orders were to be assumed legal, that it was a soldier's job to carry out any order given to him to the best of his ability.

Q: Did it tell you or talk to you or inform you anything about what occurred if you disobeyed an order by a senior officer?

A: You could be court-martialed for refusing an order, and if you refused an order in the face of the enemy, you could be sentenced to death, sir.

Q: Did they tell you anything about—did you ever hear a philosophy which might have involved the legality or illegality of orders?

William Calley Court-Martial Transcripts, National Archives Complex, College Park, Maryland, pp. 3769–834.

A: I'm not sure. I understand your question except that all orders are to be presumed legal and that all orders are to be obeyed.

Q: Well, let me ask you this. What I'm talking about and asking you about is whether or not you were given any instructions on the necessity for or whether you are required in any way, shape, or form to make a determination of the legality or illegality of an order?

A: No, sir, I was never told that I had a choice, sir.

Q: And were you told by anybody in that training what the duties were to be, what you were supposed to do in the event you received an order in connection with whether you would obey it and if so when or why or what would you do? Do you understand the question?

A: No, no, unless—

Q: If you had a doubt about an order, what were you supposed to do?

A: If I questioned the order, I was to carry it out and come back and make my complaint. That is generally what I was told. If you were given a mission to attack, you would carry it out immediately. If you had a discrepancy with the order, you reported it after you carried it out. . . .

Q: May I suggest this to you? Insofar as possible, insofar as your recollections are concerned, will you give the statements made by Captain Medina as closely as you can in his actual wording and as you understood what he said. Do I make myself clear on that?

A: Yes, sir.

Q: All right, then you go ahead.

A: He started off and he listed the men that we had lost, which was—I think, surprised everybody. Not everybody in the company had known who exactly we had lost out of other—I was quite surprised—some people in other platoons that had been lost that I had known. Then he went into that we were getting low in size. We were down about 50 percent in strength, and that the only way we would survive in South Vietnam would be to—we'd have to unite, start getting together, start fighting together, and become extremely aggressive and we couldn't afford to take anymore casualties, and that it was the people in the area that we had been operating in that had been taking the casualties on us, and that we would have to start treating them as enemy and you would have to start looking at them as enemy, that the following day we would be going into Pinkville, and Charlie Company was selected to be the main assault force and Alpha and Bravo companies would be in blocking positions. I believe he went over to the quarter-ton and somebody gave him a shovel that was in the back of the quarter-ton trailer, and he more or less drew out a sand table showing the South China Sea, the Pinkville area, My Lai (5), My Lai (6), and My Lai (4), and showed the troops where we would be coming in and explained to them that we would have to—we were going to start at My Lai (4). And we would have to neutralize My Lai (4) completely and not to let anyone get behind us, and then we would move into My Lai (5) and neutralize it and make sure there was no one left alive in My Lai (5) and so on until we got into the Pinkville area, and we would completely neutralize My Lai (5)—I mean My Lai (1) which is Pinkville. He said it was completely essential that at no time that we lose our momentum of attack because the other two companies that had assaulted the time in there before had let the enemy get behind him or he had passed through enemy, allowing him to get behind him and set up behind him, which would disorganize him when he made his final assault on Pinkville. It would disorganize him, they would lose their momentum of attack, start taking casualties, be more worried about their casualties than their mission, and that was their downfall. So it was our job to go through destroying everyone and everything in there, not letting anyone or anything get behind us and move on into Pinkville, sir. . . .

Q: All right. Was there anything in connection with that briefing of the company personnel—what was your impression as you left that meeting as to your mission and how it was to be accomplished?

A: My impression of the mission was that we would come in on a high-speed combat assault, clear My Lai (4), My Lai (5), My Lai (6), make our primary assault on Pinkville and go in there and neutralize Pinkville once and for all, sir.

Q: Was anything said in that briefing that you recollect about the people who might be left in Pinkville? If there was anybody there, did he characterize them?

A: Yes, sir.

Q: And what did he say about them?

A: That the 48th VC Battalion was in Pinkville, itself, and he said that they would be destroyed once and for all, sir.

Q: Did he make any comment about the civilians as to what they might be?

A: Well, the only remark he made to civilians, about civilians, was that a PSYWAR had prepped the area, and that the area had been completely covered by PSYWAR operations; that all civilians had left the area and that there was no civilians in the area and anyone there would be considered enemy. . . .

Q: Now I'm going to ask you this. During this operation in My Lai (4), did you intend specifically to kill any Vietnamese man, woman or child?

A: No, sir, I did not.

Q: Did you ever form any intent, specifically or generally, in connection with that My Lai operation to waste any Vietnamese man, woman or child?

A: No, sir, I didn't.

Q: I will ask you whether during that operation you at any time consciously conceived or sat down and formed an opinion to waste any man, woman or child, Vietnamese?

A: No, sir, I didn't.

Q: Now did you intend on that occasion to waste something?

A: To waste or destroy the enemy, sir.

Q: All right. Now what was your intent in connection with the carrying out of that operation as far as any premeditation or intent was concerned?

A: To go into the area and destroy the enemy that was designated there and that's it. I went in the area to destroy the enemy, sir.

Q: Well, did you form any impressions as to whether or not there were women, children or men or what was in front of you as you were going on?

A: I never sat down to analyze if they were men, women, and children. They were enemy and just people.

Q: And is that the extent of the rationalization you performed at that time?

A: Yes, sir.

Q: Did you consciously discriminate as you were operating through there so far as sex or age is concerned?

A: The only time I denoted sex was when I stopped Conti from molesting a girl. That was the only time that sex ever entered my whole scope of thinking, sir.

Q: All right. In this instance when you saw a group being supervised or guarded by Meadlo, how did you visualize that group? Did you go in to specifics in any way?

A: No, it was a group of people that were the enemy, sir.

Q: And were you motivated by other things besides the fact that those were enemy and did you have some other reasons for treating them that way all together? Now I'm talking about your briefing, and did you get any information out of that?

A: Well, I was ordered to go in there and destroy the enemy. That was my job on that day. That was the mission I was given. I did not sit down and think in terms of men, women and children. They were all classified the same, and that was the classification that we dealt with, just as enemy soldiers, sir.

Q: Who gave you that classification the last time you got it?

A: Captain Medina, sir.

Q: And where?

A: On LZ Dottie.

Q: All right. And what did he tell you in connection with that?

A: That everybody in that area would be the enemy and everyone there would be destroyed, all enemies would be destroyed.

Q: And did he classify at any of his briefings as to whether they be enemy—whether that would include men, women, and children?

A: Not specifically in that briefing. He didn't break it down and say the enemy will compose that. We had been taught that from the time we got there that men, women, and children were enemy soldiers.

Q: Was there any conversation in any other briefings which fortified that?

A: Just when someone asked a question if he meant that and he said he did, not to me directly.

Q: I'm talking about in a briefing, did he do it in a briefing?

A: Yes, sir.

Q: Now I will ask you this. Lieutenant Calley, whatever you did at My Lai on that occasion, I will ask you whether in your opinion you were acting rightly in accordance to your understanding of your directions and orders?

A: I felt then and I still do that I acted as I was directed, and I carried out the orders that I was given, and I do not feel wrong in doing so, sir.

64

PAUL MEADLO

Testimony to Peers Commission

1970

Paul Meadlo also admitted to killing noncombatants at My Lai. In the course of his testimony to the army investigators, he turned the tables and asked an investigator about his own experiences in combat.

A: Can I ask another question?

Q: Certainly.

A: Sir, during the time you were over there, did any private refuse to take orders or maybe even an NCO ever disobey any orders in combat?

Q: Let me answer that. We just had a witness in here with whom you are well acquainted. He was ordered to kill civilians at My Lai standing a few feet from them, and he refused to do it. He didn't fire a shot. He disobeyed a direct order. This is his sworn testimony.

A: From the first day we go in the service, the very first day, we are learned to take orders and not to refuse any kind of order from a non-commissioned officer or an officer would give, and that even means that if you stand on your head. An officer tells you to go and stand on your head, it's not your right to refuse that order, and you go out there and do it because you're ordered to. It's more or less in combat that it's more so there, because all though the service you hear rumors that in World War II there was people that picked up and ran, and there was officers that would shoot their men in the head to stop their people from running too. All right. You don't know what's going to come off. If you refuse the order, the son-of-a-bitch might shoot you or the next day you spend the rest of your life in the stockade for refusing an order, but you're trained to take orders from the first day you go to that damned service, and you come back and, all right, you want to try some people that had to take orders.

65

LEWIS B. PULLER JR.

From *Fortunate Son: The Autobiography of Lewis B. Puller Jr.*

1991

Lewis Puller, a U.S. marine who lost both of his legs and an arm in combat during the Vietnam War, felt that the My Lai episode stained the reputations of millions of honorable young men who had fought in Vietnam.

Lewis B. Puller Jr., *Fortunate Son: The Autobiography of Lewis B. Puller Jr.* (New York: Grove Weidenfeld, 1991), 257–58.

On November 12, 1970, at Fort Benning, Georgia, the court-martial of Lieutenant William L. Calley, Jr., for the murder of civilians at My Lai began. The trial lasted for more than four months and was the focus of such intense media coverage that it became, in effect, a forum for debate over American involvement in Vietnam. Calley was portrayed by supporters of the war as a maverick acting alone and without orders, whose actions, brought on by the stress of prolonged combat and casualties in his own unit, were an aberration from the rules of engagement. The opposing viewpoint held that his actions, if not sanctioned by higher authority, were at least tolerated and were typical of the conduct of ground units in the war.

I was deeply offended by the notion that the hideous atrocities committed by Calley and his men were commonplace in Vietnam, an inevitable consequence of an ill-advised involvement in someone else's civil war. The men I had led in combat were, like any cross section of American youth, capable of good and evil, and I felt we all were, by implication, being branded as murderers and rapists. Throughout the proceedings the reportage seemed to me to accentuate the monstrous evil of a group of men gone amok without any effort to depict fairly the discipline and courage that existed along with the forces of darkness in most units.

Lieutenant Calley was ultimately found guilty of the premeditated murder of twenty-two civilians and sentenced to life imprisonment, but I felt his punishment could never right the evil he had done or the perceptions he had helped foster of America's soldiers and marines as bloodthirsty killers. At the end of the trial I wrote letters to several local newspapers protesting that it was unfair for the Calley case to have so influenced public opinion, but the grisly photographs of murdered civilians lying in a ditch at My Lai, which had been so prominently displayed in newspapers across the country, spoke far more eloquently than my feeble words.

66

PETER STEINFELS

"Calley and the Public Conscience"

April 12, 1971

Lieutenant William Calley's conviction elicited the following comment in Commonweal *magazine.*

No one lives today in the rubble and ruins which were once Mylai. The survivors live nearby where several weeks ago they mournfully kept the anniversary of that awful morning three years ago. Hundreds of civilians were slaughtered that day. Had they been enemy soldiers, their execution in such circumstances would have still been an atrocity committed against the laws of war and military honor which our soldiers are pledged to uphold. They were, in fact, women, old men, children, babies. The village's buildings were razed, blown up, burned; the bodies of dead Vietnamese were tossed in the flames. GIs bayoneted cows. The wells were polluted with dead bodies. Mylai and its villagers were "wasted." It was not a big deal. The man who was in charge is now a national hero.

There are some good reasons not to be satisfied with the outcome of Lieutenant Calley's court martial. If Lieutenant Calley is guilty, then so are a lot of others, goes one common complaint. Calley's a scapegoat, goes another. And there is always the question of whether military justice is more military or more justice.

Whatever the truth of these allegations, the response to the Calley conviction has not stopped there. Imagine that the troops of some other nation massacre the inhabitants of a village. Several soldiers are tried and acquitted. Silence. One is charged with personally murdering 102 persons; he is convicted of murdering 22. Immediately the country is in an uproar. The capitol is flooded with protests, the convicted man with pledges of support. Regional legislatures petition for his release, and important regional officials order flags flown at half mast or other symbolic gestures of solidarity. The nation's second highest official growls that military operations "are not subject to Monday morning quarterback judgments." The chief executive seems to respond favorably to this pressure.

Peter Steinfels, "Calley and the Public Conscience," *Commonweal,* April 12, 1971, 128.

At the same time another soldier, convicted of premeditated murder of twelve villagers in a separate incident, is released after serving less than ten months in jail. One of the nation's more important politicians announces the release and the efforts under way to reinstate the soldier in the armed forces — and give him back pay.

Again — an ex-officer of an "elite" unit publicizes the fact that he did indeed commit the sensational murder he had previously denied. The government had refused to try him; it does nothing now.

If all this happened in Germany, we would declare a resurgence of Nazism. If it happened in Israel, the whole world would denounce it. If it happened in Egypt or among the Palestinians, there would be talk of the cruelest barbarism. But when it happens in America?

Mylai is no longer a village in Vietnam. It is a cancer in the conscience of America. We know what happened. Books described it. Full-color photographs were published in our national mass circulation magazines. Our President, fifteen months ago, called it "certainly a massacre."

Many Americans are simply confused, of course. They are confused on the question of "orders." They see the sum of horrors which this war is, and they rightly feel that Calley *alone* cannot be the one held guilty. The six jurors who unanimously reached their verdict were confused, too. Five of them were veterans of Vietnam combat. They felt compassion for Calley, but had no stomach for the deed. That, apparently is not the case with a large number of their fellow citizens. Explicitly or implicitly, they do not believe that what occurred at Mylai was wrong. In all the protests, the jungle paths strewn with bodies weigh very lightly.

The men, women, and infants of Mylai were slaughtered as cruelly and senselessly as were the victims of the Tate murder.[1] At Mylai, however, the victims were hundreds more; you and I armed the killers. Is this nation taking a mass "Manson murder" to its heart as an act of patriotic duty, of soldierly duty? Are our consciences that stunted, our sensitivities so shrivelled?

Much cant has been written about this or that nation being *sick:* but this time those responsible in any way for America's spiritual health had better seriously ponder the possibility.

Will the American bishops, at their forthcoming meeting, speak out without equivocation or evasion? Will they declare that certain practices revealed by the Calley testimony — e.g., marching civilians in front of a

---

[1] In 1969, actress Sharon Tate and six others were savagely murdered in Beverly Hills by members of Charles Manson's violent cult.

unit to test for land mines—cannot be condoned by *any* accepted Christian teaching on war? And that the law of God extends even to areas where the United States is conducting military operations?

Will the President truly avoid the "easy and popular course"; and, however he may deal with Calley, insist that slaughtering women and children is not an act within the compass of patriotic duty and honor?

Guilt, guilt—the pundits warn us we must not overdo the guilt. They seem to be afraid that we might actually find someone guilty. The leftish demagogues are afraid we might find the wrong people guilty.

We will not answer harder questions about the war and war crimes by deadening our abhorrence of acts like Mylai. The comment of Mylai by one brave juror is also the best judgment on the national response: "I wanted to believe it didn't happen."

67

RICHARD M. NIXON

From *RN: The Memoirs of Richard Nixon*

1978

In his memoirs, former President Richard Nixon felt the need to explain and defend his actions in arranging Lieutenant William Calley's early release from prison.

On March 29, 1971, just days after the withdrawal of ARVN troops from Laos, First Lieutenant William Calley, Jr., was found guilty by an Army court-martial of the premeditated murder of twenty-two South Vietnamese civilians. The public furor over Lam Son[2] had just begun to settle down, and now we were faced with still another Vietnam-related controversy. This one had been simmering since the fall of 1969, when the murders were first revealed.

---

[2]Lam Son 719 was the codename for a disastrous attempt by U.S. and ARVN forces to invade Laos in February 1971.

Richard Nixon, *RN: The Memoirs of Richard Nixon* (New York: Gossett and Dunlap, 1978), 449–50.

It was in March 1968, ten months before I became President, that Calley led his platoon into My Lai, a small hamlet about 100 miles northeast of Saigon. The village had been a Vietcong stronghold, and our forces had suffered many casualties trying to clear it out. Calley had his men round up the villagers and then ordered that they be shot; many were left sprawled lifeless in a drainage ditch.

Calley's crime was inexcusable. But I felt that many of the commentators and congressmen who professed outrage about My Lai were not really as interested in the moral questions raised by the Calley case as they were interested in using it to make political attacks against the Vietnam war. For one thing, they had been noticeably uncritical of North Vietnamese atrocities. In fact, the calculated and continual role that terror, murder, and massacre played in the Vietcong strategy was one of the most underreported aspects of the entire Vietnam war. Much to the discredit of the media and the antiwar activists, this side of the story was only rarely included in descriptions of Vietcong policy and practices.

On March 31 the court-martial sentenced Calley to life in prison at hard labor. Public reaction to this announcement was emotional and sharply divided. More than 5,000 telegrams arrived at the White House, running 100 to 1 in favor of clemency.

John Connally and Jerry Ford[3] recommended in strong terms that I use my powers as Commander in Chief to reduce Calley's prison time. Connally said that justice had been served by the sentence, and that now the reality of maintaining public support for the armed services and for the war had to be given primary consideration. I talked to Carl Albert and other congressional leaders. All of them agreed that emotions in Congress were running high in favor of presidential intervention.

I called Admiral Moorer on April 1 and ordered that, pending Calley's appeal, he should be released from the stockade and confined instead to his quarters on the base. When this was announced to the House of Representatives, there was a spontaneous round of applause on the floor. Reaction was particularly strong and positive in the South. George Wallace,[4] after a visit with Calley, said that I had done the right thing. Governor Jimmy Carter of Georgia said that I had made a wise decision. Two days later I had Ehrlichman announce that I would personally review the Calley case before any final sentence was carried out.

By April 1974, Calley's sentence had been reduced to ten years, with eligibility for parole as early as the end of that year. I reviewed the case

---

[3]John Connally, a conservative Democrat, was Nixon's secretary of the treasury in 1971; Gerald R. Ford was a Republican congressional representative from Michigan.
[4]George Wallace was the conservative Democratic governor of Alabama.

as I had said I would but decided not to intervene. Three months after I resigned, the Secretary of the Army decided to parole Calley.

I think most Americans understood that the My Lai massacre was not representative of our people, of the war we were fighting, or of our men who were fighting it; but from the time it first became public the whole tragic episode was used by the media and the antiwar forces to chip away at our efforts to build public support for our Vietnam objectives and policies.

68

WILLIAM C. WESTMORELAND

From *A Soldier Reports*

1976

In his memoirs, General William Westmoreland, commander of United States forces in Vietnam from 1964 to 1968, recounts his role in the investigation of My Lai.

In April 1969 the Department of the Army and a number of legislators and government officials received letters from a former soldier who had served in Vietnam, Ronald L. Ridenhour, alleging war crimes by American soldiers in an operation in March 1968, in the hamlet of My Lai (referred to as "Pinkville") in the village of Son My in Quang Ngai province. Involved was a component of the Americal Division's 11th Infantry Brigade.

Despite the obvious sincerity displayed by Ridenhour, I found it beyond belief that American soldiers, as he alleged, engaged in mass murder of unarmed South Vietnamese civilians. I directed an immediate check with MACV headquarters in Saigon. When the MACV inspector general reported that something untoward might have occurred, the Inspector General of the Department of the Army began an immediate investigation, which was subsequently pursued by the Army Criminal Investigation Division. It resulted in charges against four officers and nine

William C. Westmoreland, *A Soldier Reports* (New York: Doubleday, 1976), 375–80.

enlisted men and trials of two officers and three enlisted men. Twenty-five former enlisted men were implicated, but since they had already been discharged from the Army, they were beyond the Army's jurisdiction.

Almost as deplorable as the events alleged was the possibility that officers of the 11th Brigade and the Americal Division had either covered up the incident or failed to make a comprehensive investigation. The developing evidence in the criminal investigation and the indications of command dereliction led [Army] Secretary Resor and me to arrange for an additional formal inquiry into the adequacy of the criminal investigation and the possible suppression of information. When I learned that some members of President Nixon's administration wanted to white-wash any possible negligence within the chain of command, I threatened through a White House official to exercise my prerogative as a member of the Joint Chiefs of Staff to go personally to the President and object. That squelched any further pressure for whitewash.

My first thought was to propose a civilian commission to enhance the credibility of the findings, but upon reflection I decided that the situation in Vietnam was so complex, the terminology and experience so alien to civilians, that it would be better to have a board headed by a military man assisted by civilian lawyers. To head the board the Secretary and I selected the former I Field Force commander, General Peers, who had a reputation throughout the Army for objectivity and fairness. Ray Peers had also been a division commander in Vietnam and thus was thoroughly familiar with conditions; he had never had jurisdiction over any activity in Quang Ngai province. Because he had entered the Army through ROTC at the University of California at Los Angeles, there could be no presumption that ties among brother officers from West Point would be involved. For the civilian legal counsel we obtained the services of two distinguished New York attorneys, Robert MacCrate and Jerome K. Walsh, Jr., who assisted General Peers but reported directly to Secretary Resor.

As a result of evidence developed by the Peers board, charges were brought against twelve officers, primarily involving dereliction of duty in suppressing information and failing to obey lawful regulations. These included the former Americal Division commander, General Koster, who at the time of the investigation was Superintendent of the Military Academy. Lest any findings reflect adversely on the Academy, he requested relief from that post.

Under usual courts-martial practice, the pretrial investigations would have been assigned to the Army commanders in whose commands the officers were serving. To simplify procedures and assure that all would be judged by the same criteria, I instead transferred all the officers associated with charges stemming from the Peers investigation to the First Army and assigned the investigation to its commander, Jack Seaman. I am sure it was for him a demanding assignment. After detailed review of the Peers board findings and further investigation, Seaman concluded that evidence was insufficient to bring any of the officers to trial for dereliction of duty except the former 11th Infantry Brigade commander, who was subsequently court-martialed and acquitted after a lengthy trial.

Even though the evidence as reviewed by a man of honesty and courage proved insufficient for trial or conviction, something had to be remiss in the American Division's chain of command if anything so reprehensible and colossal as the My Lai massacre occurred without some responsible official either knowing or at least suspecting.

It was true that at President Johnson's direction the 11th Infantry Brigade had been deployed to Vietnam before completing its training, in order to get the troops there in advance of an arrangement the President was hoping—vainly, as it turned out—to achieve with the North Vietnamese for a cease-fire and a freeze in troop strength. Although I committed the brigade in a quiet sector so training might continue, just over a month later the troops were caught up in the enemy's Tet offensive. It was also true that the unit at My Lai was part of a temporary or *ad hoc* task force lacking the unity of an established organization. Yet those were no more than mitigating factors, not excuses. Although the division commander did order an investigation, he made a basic error in assigning the investigation to the commander of the responsible unit, the 11th Infantry Brigade.

Contrary to my standing directive, not even the fact that an investigation, however perfunctory, took place was reported either to the intermediate headquarters, the III Marine Amphibious Force and the United States Army, Vietnam, or to my headquarters; and reference to my records reveals that I visited the American Division and the 11th Infantry Brigade on April 20, 1968, only a few weeks after the massacre, and nobody intimated to me that anything was under suspicion or even remotely remiss except a Red Cross official who complained about mail service.

As with civilian justice, the United States Army is committed to due legal process under which an individual must be presumed innocent until proven guilty. Indeed, the Uniform Code of Military Justice gives the accused extraordinary protection. The Army cannot bring a man to trial simply on the basis of unsubstantiated allegations, however plausible, however widely publicized. If pretrial investigation fails to produce substantiated evidence sufficient to warrant trial, the accused must be protected against prejudicial statements made during the investigation, for the investigation itself is no trial. Thus the full details of the Peers board, even as the proceedings of a grand jury, have been kept secret.

The U. S. Army does have another pretrial procedure not usually available to civilian authority: administrative review to determine if the performance of the person under investigation conformed to established standards of the military profession. If the charges against him are dropped, it means only that further criminal proceedings are unwarranted, not that his performance has been found adequate. As with the court-martial system, administrative review is not the province of the Chief of Staff but of the Secretary of the Army, although the Secretary has the benefit of the Chief of Staff's recommendations. As a result of administrative review, Secretary Resor took administrative action against two general officers of the Americal Division and eight others whose performances in connection with the investigation and reporting of My Lai were deemed to be below the standards expected of individuals of their positions, grades, and experiences.

In the criminal cases, acquittal resulted in all but that of a platoon leader, First Lieutenant William L. Calley, Jr. Charged with the murder of more than a hundred civilians, he was convicted on March 29, 1971, of the murder of "at least" twenty-two. He was sentenced to dismissal from the service and confinement at hard labor for life, but the latter was reduced by judicial review to twenty years and further reduced after my retirement by Secretary of the Army Howard Callaway to ten years, an action that President Nixon sustained. The case was subsequently and for a long time under judicial appeal in the federal courts.

Lieutenant Calley was legally judged by a jury whose members all were familiar with the nature of combat in Vietnam and well aware that even the kind of war waged in Vietnam is no license for murder. The vast majority of Americans in Vietnam did their best to protect civilian lives and property, often at their own peril. That some civilians, even many, died by accident or inevitably in the course of essential military operations dictated by the enemy's presence among the people was no justification or rationale for the conscious massacre of defenseless babies, children, mothers, and old men in a kind of diabolical slow-motion nightmare that went on for the better part of a day, with a cold-blooded break for lunch. I said at the time of the revelation: "It could not have happened—but it did."

Although I can in no way condone Lieutenant Calley's acts—or those of any of his colleagues who may have participated but went unpunished—I must have compassion for him. Judging from the events at My Lai, being an officer in the United States Army exceeded Lieutenant Cal-

ley's abilities. Had it not been for educational draft deferments, which pre-
vented the Army from drawing upon the intellectual segment of society
for its junior officers, Calley probably never would have been an officer.
Denied that usual reservoir of talent, the Army had to lower its standards.
Although some who became officers under those conditions performed
well, others, such as Calley, failed.

An army has a corps of officers to insure leadership: to see that orders
are given and carried out and that the men conduct themselves properly.
Setting aside the crime involved, Lieutenant Calley's obvious lack of
supervision and failure to set a proper example himself were contrary to
orders and policy, and the supervision he exercised fell far short.

In reducing standards for officers, both the United States Army and
the House Armed Services Committee, which originated the policy of
deferments for college students, must bear the blame. It would have
been better to have gone short of officers than to have accepted applicants
whose credentials left a question as to their potential as leaders.

Some of the public sympathy that developed for Calley may be attributed to a rash of intemperate allegations that followed the revelation of
the My Lai massacre. If Calley was guilty, why not also his superiors,
including Westmoreland? Citing the Nuremberg and Yamashita trials[5] of
the World War II era as precedents, the chief proponent of such a concept, Telford Taylor, professor of law at Columbia University, who had
been chief counsel for the prosecution at Nuremberg, pondered whether
not only Westmoreland but also civilian officials in Washington should
be tried for war crimes. On a television talk show promoting his book
Nuremberg and Vietnam: An American Tragedy, he said that if the same
standards had been applied to the My Lai trial that had been in the trial
of General Tomoyuki Yamashita, "there would be a very strong possibility
that they [myself and civilian government officials] would come to the
same end as he did."

His was an emotional outburst. Many a jurist as schooled in the law
as Professor Taylor responded that the critics were ignoring two cardi-
nal principles of the Nuremberg and Yamashita cases: intent and efforts
to prevent war crimes. It was declared at Nuremberg, for example, that
in order to establish a commander's criminal liability for atrocities there
had to be ". . . a personal neglect amounting to wanton, immoral disregard of the actions of subordinates amounting to acquiescence." In any

---

[5]The Nuremberg war crimes trials in 1946 convicted many high-ranking Nazi leaders
for crimes committed by Nazis throughout World War II. Twelve were sentenced to death.
Also following World War II, Japanese General Tomoyuki Yamashita was convicted of
atrocities committed by his troops and was hanged in 1946.

event, while lamenting My Lai and any other war crime or felony with every fiber of my being, I am convinced that my actions in Vietnam and the efforts I made to forestall the kind of thing that happened at My Lai will stand every moral and legal test before both the bar of justice and the court of history.

During the investigations of the events at My Lai, it came out that within a few days of the action I had forwarded the 11th Infantry Brigade a commendation. It was based on the brigade's official report of 128 enemy killed at My Lai and four[6] weapons captured. Why had such a disparity between killed and weapons failed to alert me that something untoward might have occurred?

The report on My Lai that reached my headquarters, where it was accepted in good faith, attracted no special attention for two reasons. As opposed to warfare against the enemy's big units, high body counts and low numbers of weapons collected in the war against the guerrilla in hamlet and village were not uncommon (the dead were presumed to be armed combatants, not civilians). To assure accurate reporting, I had had several reports like that investigated; the investigations revealed that guerrillas were adept at disposing of weapons in paddy or canal, and many guerrillas often were armed only with grenades and explosives. Secondly, in keeping with my desire to reward men and units for good performances, it was a practice in my headquarters either for me or my staff to select from among the volume of daily field reports those that appeared particularly noteworthy, whereupon numerous routine commendations would be prepared for my signature. Having passed up the chain of command, the reports gained added credibility from at least implicit endorsement of the intermediate headquarters: company, battalion, brigade, division, field force, and United States Army, Vietnam. We had to rely on the presumed and generally established veracity of the reports and the chain of command.

Despite excellent communications and speedy transport, no one person could know everything that happened in Vietnam nor be at the scene of every action. Any senior commander has to depend upon subordinates — especially commissioned officers — for local supervision. In naming General Koster the commander of the Americal Division I had acted not from personal knowledge of him but from the recommendation of the Army's Chief of Staff, General Johnson, and my deputy, General Abrams. I nevertheless had no reason to question Koster's ability — nor that of his subordinate commanders — to control his troops and comply with my

---

[6]The official report stated three, not four, weapons.

regulations on reporting irregularities that might be revealed following a routine survey of the battlefield, a normal duty of a field commander. Although the press on several occasions alerted me to events warranting inquiry, not so on My Lai.

Over the years a number of other battlefield irregularities, some of which were war crimes, were reported or alleged. The Army investigated every case, no matter who made the allegation, with professionally qualified, noninvolved parties. Some investigations resulted in disciplinary administrative action, some in courts-martial. Yet none of the crimes even remotely approached the magnitude and horror of My Lai, and many of the allegations, principally those leveled by individuals testifying under the aegis of such organizations as the Vietnam Veterans Against the War and a so-called Citizens Commission of Inquiry into U. S. War Crimes in Vietnam, were backed by no responsible evidence.

Glossary

Here are definitions of some acronyms and abbreviations that you will encounter in the introduction and documents.

Americal   The nickname for the 23rd Infantry Division

AO   Area of operations

ARVN   Army of the Republic of Vietnam (South Vietnamese army)

Avn Co   Aviation company

BG   Brigadier general

CH-47   A U.S. army helicopter that delivered troops into battle

Charlie   Slang for Vietcong

CIA   Captured in action

CID   Criminal Investigation Division, U.S. army

Co   Company

CO   Commanding Officer

Col. or **COL**   Colonel

CP   Command post

CPT   Captain

DMZ   Demilitarized zone (the border between North Vietnam and South Vietnam)

DRV   Democratic Republic of Vietnam (North Vietnam)

1LT   First lieutenant

FOB   Forward operating base (a battalion-level command post in the field)

free-fire zone   Any area in which permission was not required from American commanders prior to firing on targets. In theory, at least, U.S. forces had already relocated all civilians; anyone left behind was assumed to be Vietcong.

GI   Nickname for a U.S. soldier

GVN   Government of Vietnam (South Vietnam)

hooch or **hootch**   U.S. military slang for a Vietnamese home

Huey or **HUI** or **HUID**   Nickname for a variety of U.S. army helicopters

IG   Inspector general

KHA   Killed in hostile action (a euphemism for U.S. troops killed in combat). The usual KIA designation was not used because Vietnam was not a declared war.

KIA   Killed in action

LF   Local force, a Vietcong military unit subordinate to a district or province headquarters of the National Liberation Front (NLF)

LRP or **LRRP**   "Lurp," or long-range reconnaissance patrol

Lt or **LT**   Lieutenant

LTC   Lieutenant colonel

LZ   Landing zone (the area in which a helicopter lands to deliver or pick up soldiers)

M-16   The primary rifle carried by American soldiers in Vietnam

M-60   A portable machine gun used by American troops in Vietnam

MACV   Military Assistance Command, Vietnam (the headquarters of U.S. military operations in Vietnam)

MF   Main force (regular Vietcong and North Vietnamese soldiers)

MG   Major general

MI   Military intelligence

NCO   Noncommissioned officer

NLF   National Liberation Front (the political arm of the Vietcong)

NVA   North Vietnamese Army

OH-23   A U.S. army helicopter

Pfc.   Private, first class

psyops   Psychological operations

Plt   Platoon

Psywar or **PSYWAR**   Psychological warfare. U.S. troops engaged in psychological warfare by distributing information and propaganda to South Vietnamese civilians, Vietcong, and North Vietnamese soldiers.

PW   Prisoner of war

R and R   Rest and recreation, vacations given to U.S. soldiers midway through their tour of duty

RTO   A U.S. army radio operator

RVN   Republic of Vietnam (South Vietnam)

RVNAF   RVN (South Vietnam) armed forces

Sgt. or **SGT**   Sergeant

SOP   Standard operating procedure

Stars and Stripes   A U.S. military newspaper

Tet   Vietnamese New Year

TF   Task force

tunnel rat   An American soldier trained to enter underground tunnels in search of enemy soldiers

VC   Vietcong

VCS   Vietcong sympathizers

WHA   Wounded in hostile action (a euphemism for U.S. troops wounded in combat). The usual WIA designation was not used because Vietnam was not a declared war.

WIA   Wounded in action

A My Lai Chronology
(1967–1974)

1967

September: General William Westmoreland activates the 23rd Infantry Division, the Americal Division.

December: Charlie Company of the 11th Infantry Brigade is deployed to Quang Ngai province, South Vietnam.

1968

January: Task Force Barker is formed.

January 31: The Vietcong and North Vietnamese launch the Tet Offensive.

March 16: The massacre takes place at My Lai in Quang Ngai province.

April: Ronald Ridenhour first hears of the My Lai massacre.

1969

March 29: Ronald Ridenhour posts his letter asking for a formal investigation of My Lai.

April: The U.S. army asks Colonel William Wilson to make a preliminary investigation of the My Lai incident.

September 5: Lieutenant William Calley is charged with murder.

November 13: Seymour Hersh's article exposing the massacre is picked up by wire services and published in newspapers throughout the United States. Robert Smith's article exposing the massacre is published in the *New York Times*.

November 26: The U.S. army instructs Lieutenant General William Peers to conduct an official investigation into the massacre and its cover-up.

December 5: *Life* magazine publishes Ronald Haeberle's color photographs of the My Lai massacre.

1970

March: Captain Ernest Medina is charged with murder. The Peers Commission makes its formal report, charging twenty-five men with murder, conspiracy, and/or perjury. Charges are dropped against all but four; William Calley and Ernest Medina are charged with murder. Colonel Oran K. Henderson is charged with dereliction of duty, failure to report a war crime, and perjury. Major General Samuel Koster is charged with dereliction of duty and failure to report civilian casualties.

September: Ernest Medina is acquitted of all charges.

November 10:   William Calley's court-martial begins.

1971

March 29:   William Calley is convicted of multiple acts of murder and sentenced to life in prison at hard labor. Samuel Koster is acquitted of all criminal charges, but the secretary of the army reduces his rank to brigadier general and strips him of his Distinguished Service Medal.

August: William Calley's life sentence is reduced to twenty years.

November: Oran K. Henderson is acquitted of all criminal charges.

1974

April: William Calley's twenty-year sentence is reduced to ten years.

November: Secretary of the Army Howard Calloway officially pardons William Calley.

Questions for Consideration

1. According to international law, how are soldiers required to treat civilian noncombatants?
2. How did the nature of the war in Vietnam make civilian casualties more common than in other American wars?
3. Describe the training in the rules of land warfare and international law concerning noncombatants given to the troops in Charlie Company.
4. Was the training adequate? Why or why not?
5. As you read the testimony of the members of Charlie Company, how would you characterize their attitudes toward Vietnamese people?
6. In your opinion, what was the mood of Charlie Company's troops at the briefings on the evening of March 15, 1968? Why?
7. What do you think Captain Ernest Medina told the soldiers at the March 15 briefing about the next day's operation? Did he clearly tell them to "waste" all men, women, and children in the village?
8. What type of soldier was Captain Ernest Medina? How would you explain his actions on March 16, 1968?
9. What type of soldier and person was Lieutenant William Calley? How had the rhetoric of the cold war and anticommunism shaped his own thinking about the enemy in Vietnam?
10. In your opinion, why did some soldiers in Charlie Company refuse to obey Calley's order to shoot civilians while others readily complied?
11. Were the troops of Charlie Company guilty of war crimes on March 16, 1968?
12. What about the U.S. soldiers "finishing off" wounded civilians who would not have survived anyway? Could this action be considered a "mercy killing"?
13. Is "mercy killing" a war crime? Why or why not?
14. How would you explain the multiple incidents of rape that occurred at My Lai during the course of the massacre?
15. None of the men in Charlie Company who refused Calley's orders to shoot civilians ever reported the incident to higher authorities, even though they believed that a war crime had occurred. Why?
16. Why was it a nonmember of the unit—helicopter pilot Hugh Thompson—who reported the massacre?

17. Why did commanding officers in the Americal Division fail to investigate the atrocity thoroughly?
18. Why were so few of the participants in the My Lai massacre punished for their participation?
19. What does the My Lai massacre and its aftermath tell you about the Vietnam War?
20. In your opinion, was My Lai an aberration during the Vietnam War or a relatively common occurrence? Why?
21. Why should Americans today still be interested in what happened at My Lai on March 16, 1968?

Suggested Readings

Long before dozens of newspapers published Seymour Hersh's My Lai exposé in November 1969, critics of the American war in Vietnam had raised the issue of war crimes and morality. A number of European intellectuals, who were convinced that United States policymakers had confused communism with nationalism in Southeast Asia, questioned the use of so much military firepower to deal with what was essentially a political issue. The United States war machine was too powerful and too careless, these critics contended, and in the process hundreds of thousands of Vietnamese peasants, in both South Vietnam and North Vietnam, were being killed, wounded, or made homeless. Jean-Paul Sartre, the legendary French existentialist philosopher, accused the United States in his book *On Genocide* (Boston: Beacon Press, 1968) of conducting a genocidal war against the Vietnamese people. British philosopher Bertrand Russell, in *War Crimes in Vietnam* (London: George Allen and Unwin, 1967), had already leveled similar charges at the United States.

In 1965–66, Sartre, Russell, and a host of other European intellectuals convened a mock war crimes trial in which antiwar representatives from around the world accused the United States of violating every article of international law and engaging in war crimes in Vietnam. John Duffet edited the proceedings and published them as *Against the Crime of Silence: Proceedings of the International War Crimes Tribunal* (New York: Clarion, 1970). A number of American writers echoed the sentiments of the mock International War Crimes Tribunal, including Telford Taylor, *Nuremberg and Vietnam: An American Tragedy* (Chicago: Quadrangle, 1970); Robert F. Drinan, *Vietnam and Armageddon: Peace, War, and the Christian Conscience* (New York: Sheed and Ward, 1970); Erwin Knoll and Judith Nie McFadden, eds., *War Crimes and the American Conscience* (New York: Holt, Rinehart, and Winston, 1970); and Seymour Melman, *In the Name of America: The Conduct of the War in Vietnam by the Armed Forces of the United States as Shown by Published Reports, Compared with the Laws of War Binding on the United States Government and Its Citizens* (New York: Clergy and Laymen Concerned about Vietnam, 1968).

The articles by Seymour Hersh and Robert Smith in November 1969, along with *Life* magazine's publication of Ron Haeberle's grisly My Lai photographs, unleashed a torrent of criticism and national soul-searching. Hersh was first out of the publishing starting blocks with his book *My Lai 4: A Report on the Massacre and Its Aftermath* (New York: Random House, 1970). Other accounts soon followed, including Martin Gershin, *Destroy or Die: The Story of Mylai* (New Rochelle: Arlington House, 1971), and Richard Hammer, *One Morning in the War: The Tragedy of Son My* (New York: Coward-McCann, 1970).

Books about William Calley and Ernest Medina quickly found their way into bookstores as well. Virtually all of them concluded that Medina had been an exemplary officer up to the day of the massacre and that Calley was a quite ordinary American young man. John Sack's *Lieutenant Calley: His Own Story* (New York: Viking, 1970) was designed to generate royalties for Calley's legal defense fund. The best look at Medina is Mary McCarthy's *Medina* (New York: Harcourt Brace Jovanovich, 1972). Also see Richard Hammer, *The Court-Martial of Lt. Calley* (New York: Coward-McCann, 1971); Wayne Greenhaw, *The Making of a Hero: The Story of Lieut. William Calley Jr.* (Louisville: Touchstone, 1971); and Arthur Everett, Kathryn Johnson, and Harry F. Rosenthal, *Calley* (New York: Dell, 1971).

Revelations about My Lai made their way into the headlines just before the controversy over the Pentagon Papers burst onto the national scene. When the *New York Times* and the *Washington Post* published the so-called Pentagon Papers, it became abundantly clear that top government officials, for more than a decade, had systematically misled the public about U.S. policy in Vietnam. The fact that there had been a cover-up of the My Lai massacre came as little or no surprise. A number of books detailed the conspiracy, including Seymour M. Hersh, *Cover-Up: The Army's Secret Investigation of the Massacre at My Lai* (New York: Random House, 1972), and Richard A. Falk, Gabriel Kolko, and Robert Jay Lifton, eds., *Crimes of War* (New York: Random House, 1971). Also see Joseph Goldstein, Burke Marshall, and Jack Schwartz, *The My Lai Massacre and Its Cover-up: Beyond the Reach of Law?* (New York: Free Press, 1976). The conclusions of the Peers Commission were published as *The My Lai Inquiry* (New York: Norton, 1979).

During the courts-martial of Calley, Medina, Henderson, and Koster, Americans debated the morality of war in general and the Vietnam War in particular. When the Vietnam Veterans Against the War convened the Winter Soldier Investigation early in 1971, a parade of former soldiers testified of the wholesale rape, torture, and murder of noncombatants in Vietnam. A number of books appeared examining the issue of war crimes. Among the most controversial were Vietnam Veterans Against the War, *The Winter Soldier Investigation: An Inquiry into American War Crimes* (Boston: Beacon Press, 1972); Citizens Commission of Inquiry, *Dellums Committee Hearings on War Crimes in Vietnam* (New York: Vintage, 1972); and T. J. Farer, *The Laws of War 25 Years after Nuremberg* (New York: Carnegie Endowment for International Peace, 1971). Also see Sheldon M. Cohen, *Arms and Judgement:*

Law, Morality, and the Conduct of War in the Twentieth Century (Boulder: Westview, 1989); Gabriel Kolko, *Anatomy of a War: Vietnam, the United States, and the Modern Historical Experience* (New York: Pantheon, 1986); Leo Kuper, *Genocide: Its Political Use in the Twentieth Century* (New Haven: Yale University Press, 1981).

Within the past decade, three books have shed new light on the My Lai massacre and the events leading up to it. Neil Sheehan's *A Bright Shining Lie* (New York: Random House, 1988) put the My Lai massacre into a macabre context by arguing that Charlie Company certainly murdered hundreds of noncombatants on March 16, 1968, but that during the previous two years, indiscriminate aerial and artillery bombardment had killed up to fifty thousand civilians in Quang Ngai province. Christian Appy's *Working-Class War* (Chapel Hill: University of North Carolina Press, 1993) described just how frustrating it was for American GIs to sustain so many wounds from booby traps and snipers and how it made all Vietnamese suspect in their eyes. Finally, Michael Bilton and Kevin Sim's *Four Hours in My Lai* (New York: Penguin, 1992) returns to My Lai nearly a quarter of a century after the massacre. The authors interviewed members of Charlie Company as well as a number of Vietnamese survivors, reconstructing the entire episode.

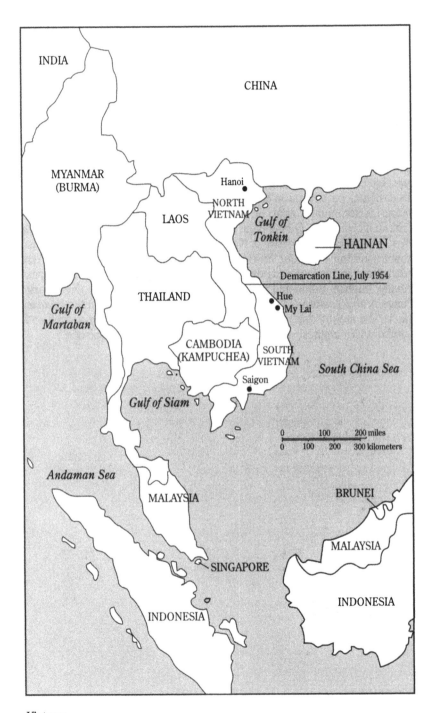

Vietnam.

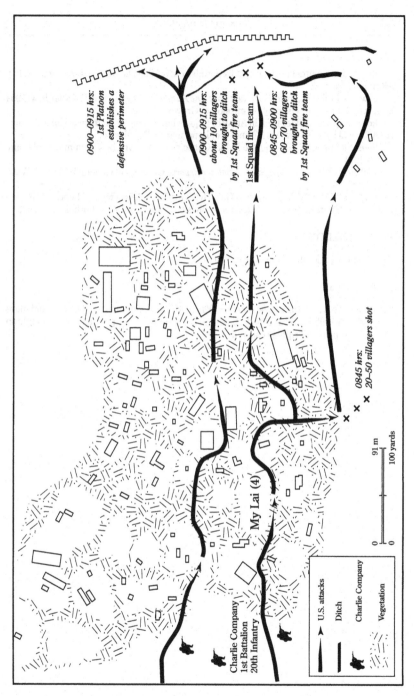

The My Lai Massacre, March 16, 1968.

(Continued from p. iv)

"Notes and Comment." Reprinted by permission; © 1969 *The New Yorker* Magazine, Inc. All rights reserved.

Lewis B. Puller Jr., from *Fortunate Son: The Autobiography of Lewis B. Puller Jr.* © 1991 by Lewis B. Puller Jr. Used by permission of Grove/Atlantic.

Robert Smith, "Officer Kept in Army in Inquiry Into Killing of Vietnam Civilians." Copyright © 1969 by The New York Times Co. Reprinted by Permission.

Peter Steinfels, "Calley and the Public Conscience." Copyright © Commonweal Foundation 1971.

William C. Westmoreland, from *A Soldier Reports.* Courtesy of General William C. Westmoreland.

William Wilson, "I Had Prayed to God That This Thing Was Fiction." Reprinted by permission of *American Heritage Magazine,* a division of Forbes, Inc. © Forbes, Inc., 1990.

PHOTO CREDITS

Ron Haeberle, Life Magazine. © Time Inc.

MAP CREDITS

Map from *Historical Atlas of the Vietnam War* by Harry G. Summers Jr. Designs and maps copyright © 1995 by Swanston Publishing Ltd. Reprinted by permission of Houghton Mifflen Company. All rights reserved.

Index

Printed in the United States
By Bookmasters